T0384363

Solar Energy: Advancements and Challenges

RIVER PUBLISHERS SERIES IN ENERGY SUSTAINABILITY AND EFFICIENCY

Series Editors

PEDRAM ASEF
Lecturer (Asst. Prof.) in Automotive Engineering,
University of Hertfordshire,
UK

The "River Publishers Series in Sustainability and Efficiency" is a series of comprehensive academic and professional books which focus on theory and applications in sustainable and efficient energy solutions. The books serve as a multi-disciplinary resource linking sustainable energy and society, fulfilling the rapidly growing worldwide interest in energy solutions. All fields of possible sustainable energy solutions and applications are addressed, not only from a technical point of view, but also from economic, social, political, and financial aspects. Books published in the series include research monographs, edited volumes, handbooks and textbooks. They provide professionals, researchers, educators, and advanced students in the field with an invaluable insight into the latest research and developments.

Topics covered in the series include, but are not limited to:

- Sustainable energy development and management;
- Alternate and renewable energies;
- Energy conservation;
- Energy efficiency;
- Carbon reduction;
- Environment.

For a list of other books in this series, visit www.riverpublishers.com

Solar Energy: Advancements and Challenges

Editors

Gaurav Saini
Department of Mechanical Engineering,
Harcourt Butler Technical University Kanpur, India

Korhan Cengiz
Faculty of Engineering, Trakya University, Turkey

Sesha Srinivasan
Department of Engineering Physics,
Florida Polytechnique University, USA

Sanjeevikumar Padmanaban
Department of Business Development and Technology,
CTIF Global Capsule (CGC) Laboratory, Aarhus University, Denmark

Krishna Kumar
Research and Development Unit, UJVN Ltd., Dehradun,
Uttarakhand, India

Published 2023 by River Publishers
River Publishers
Alsbjergvej 10, 9260 Gistrup, Denmark
www.riverpublishers.com

Distributed exclusively by Routledge
4 Park Square, Milton Park, Abingdon, Oxon OX14 4RN
605 Third Avenue, New York, NY 10017, USA

Solar Energy: Advancements and Challenges / Gaurav Saini, Korhan Cengiz, Sesha Srinivasan, Sanjeevikumar Padmanaban and Krishna Kumar.

Routledge is an imprint of the Taylor & Francis Group, an informa business

ISBN 978-87-7022-703-2 (print)
ISBN 978-10-0084-733-8 (online)
ISBN 978-1-003-37390-2 (ebook master)

Contents

Preface

Energy is a key source of economic growth due to its involvement as the primary input. Energy drives economic productivity and industrial growth. It can be considered as the prime requirement for the modern economy. For centuries, the energy need has been fulfilled using conventional fuels such as coal, petroleum, and other natural-gas-based fossil fuels. However, in the last few years, the awareness has been increased regarding the climate, and attention has now been shifted toward green and renewable sources of energy. Renewable energy is energy from naturally replenishing sources. Renewable resources are virtually inexhaustible in duration but limited in the amount of energy available per unit of time.

Solar energy is a renewable source of energy which can be used to produce heat or generate electricity. The total amount of solar energy available on Earth's surface is vastly in excess of the world's current and anticipated energy requirements. In the 21st century, solar energy is expected to become increasingly attractive as a renewable energy source. An increase in the share of solar energy may unstabilize the grid. To overcome the issues of grid instability, specifically in remote areas, BIM- and GIS-based microgrid planning based on data can be effectively used. BIM and GIS are used to assess alternative solutions and big data analytics in building solar electrical systems according to the planning requirements and managing assets. The integration of BIM and GIS information systems for microgrid planning is appealing due to its potential benefits, such as it decreases the microgrid planning time and cost.

The present book is about the advancements in the technology for harnessing solar energy and the challenges associated with different modes of utilizing this inexhaustible renewable energy source.

Chapter 1 explains the solar energy and fabrication of crystalline silicon in this book. Chapter 2 presents the uncertainty-based battery sizing in distributed renewable systems. Chapter 3 elucidates about design and development of a solar-powered hybrid energy bank. Chapter 4 established an ANN-based maximum power point tracking of a PV system. Chapter 5 models the BIM- and GIS-based residential microgrid. Chapter 6 investigates the thermo-hydraulic performance of corrugated and impinging jet solar air heaters. Chapter 7 explains the leakage current in solar photovoltaic modules. Chapter 8 analyzes the reliability and degradation of the crystalline silicon photovoltaic module. Chapter 9 synthesizes and characterized the botanical dye-sensitized solar cell (DSSC) based on TiO_2 using Capsicum Annuum and Coriandrum Sativum extracts.

This book shall be helpful for the researchers, academicians, technologists, innovators, and industry experts working in the area of renewable energy, artificial intelligence, micro-grid, water resource, and energy trading/auditing.

List of Contributors

Bahrein, W. M., *Department of Electrical and Electronics Engineering, Universiti Teknologi Petronas, Malaysia*

Devamurugan, *Department of Electrical and Electronics Engineering, Universiti Teknologi Petronas, Malaysia*

Farooq, J., *School of Engineering, University of Petroleum & Energy Studies, India*

Garimella, R. C., *Associate Professor, Department of Electrical & Electronics Engineering, Methodist College of Engineering and Technology, India*

Gupta, M., *JECRC University, India*

Gupta, R., *Department of Energy Science and Engineering, Indian Institute of Technology Bombay, India*

Kannan, R., *Department of Electrical and Electronics Engineering, Universiti Teknologi Petronas, Malaysia*

Kuey, T. C., *Department of Electrical and Electronics Engineering, Universiti Teknologi Petronas, Malaysia*

Kumar, K., *Research Scholar, IIT Roorkee, India*

Kumar, M., *Institute for Energy Technology Kjeller, Norway*

Kumar, M., *Methodist College of Engineering and Technology, India*

Kumar, R., *Department of Energy Science and Engineering, Indian Institute of Technology Bombay, India*

Kumar, R. S., *Department of Electrical and Computer Engineering, King Abdulaziz University, Saudi Arabia*

Madeti, S. R. K., *Assistant Professor, University of Santiago de Chile, Chile*

Maity, S., *School of Advanced Materials, Green Energy and Sensor Systems, Indian Institute of Engineering Science and Technology, India*

Meena, R., *Department of Energy Science and Engineering, Indian Institute of Technology Bombay, India*

Nayak, K. R., *Methodist College of Engineering and Technology, India*

Pachauri, R. K., *School of Engineering, University of Petroleum & Energy Studies, India*

Panda, T., *School of Advanced Materials, Green Energy and Sensor Systems, Indian Institute of Engineering Science and Technology, India*

Saini, G., *Assistant Professor, School of Advanced Materials, Green Energy and Sensor Systems, Indian Institute of Engineering Science and Technology Shibpur, India*

Saini, R. P., *Professor, HRED, IIT Roorkee, India*

Shankar, T. B., *Methodist College of Engineering and Technology, India*

Shrivastava, A., *Poornima University, Jaipur, India*

Shrivastava, R., *Vivekananda Global University, Jaipur, India*

Singh, R., *Department of Electrical and Electronics Engineering, Universiti Teknologi Petronas, Malaysia*

Yadav, S., *Research Scholar, HRED, IIT Roorkee, India*

Zhou, Y., *Sustainable Energy and Environment Thrust, Function Hub, The Hong Kong University of Science and Technology, China*
Department of Mechanical and Aerospace Engineering, The Hong Kong University of Science and Technology, Clear Water Bay, China

List of Figures

List of Tables

List of Notations and Abbreviations

Al Aluminum
c-Si Crystalline silicon
D&D Defects and degradations
EA Activation energy (kJ mol–1)
EVA Ethylene-vinyl acetate
HVS High voltage stress
I Cell output current (A)
Impp Current at maximum power point (A)
Isc Short circuit current (A)
Jo Saturation current density (A m–2)
K Boltzmann constant (J K–1)
Na Sodium
n Ideality factor
PA Polyamide
PET Polyethylene terephthalate
PID Potential induced degradation
Pmpp Power at maximum power point (W)
PVDF Polyvinylidene fluoride
PVF Polyvinyl fluoride
PV Photovoltaic
Rs Cell series resistance (Ω)
Rsh Cell shunt resistance (Ω)
Si Silicon
T Operating temperature (K)
UV Ultraviolet
V Cell output voltage (V)
Vmpp Voltage at maximum power point (V)
Voc Open circuit voltage (V)
σ Conductivity (Ω–1)

Chapter 1

Fabrication of Crystalline-Silicon Based Solar Cell

Tamalika Panda and *Santanu Maity

School of Advanced Materials, Green Energy and Sensor Systems, Indian Institute of Engineering Science and Technology, India
*Corresponding author: maitys.iiest@gmail.com

Abstract

Motivation: With the increase in world population, fossil-fuel-based energy will last for the next few years. Therefore, there is a need to shift from fossil-fuel-based energy sources to renewable energy sources. Solar cells directly convert sunlight into electrical energy. But the high cost is one of the major challenges of power generation by solar cells. The cost of a solar cell depends on different parameters like material consumption, silver consumption, specific fabrication facility, etc. There are several ways to mitigate the cost of solar cell and module production. Material cost can be optimized by the thinning of wafer thickness, whereas optimization of grid parameters can reduce the silver consumption. On the other hand, one of the important cost reduction parameters is the fabrication cost. This chapter describes the fabrication technology of solar cells in detail. Various steps are involved in solar cell fabrication, such as texturing, emitter diffusion, PSG removal and edge isolation, ARC coating, metallization and screen printing, and testing. CZ/FZ wafer is used to fabricate the solar cell. After the saw damage removal, the surface of the wafers shows high optical reflectivity, which can be reduced by texturing the surfaces and followed by the ARC coating.

The next step is to create the junction, which can be done by emitter diffusion, wherein the case of p-type c-Si substrate n-type top layer will be the emitter and vice versa. To disconnect the top and bottom surface of the solar cell electrically, edge isolation is done. After that, passivation is done to passivate the surface defects by depositing a thin dielectric layer. For electrical connections, metallization is done to make the contacts in the form of fingers and busbars. For front contact, silver (Ag), and for rear contact, aluminum (Al) is used, and it is done by screen printing followed by co-firing. One of the important issues to reduce the solar cell cost is to increase the solar cell efficiency by minimizing the power losses (electrical and optical power losses) in the front grid structure optimization in the solar cell.

1.1 Introduction

According to various reports, it has been found that the fossil-fuel-based energy sources (coals and petroleum) are now reduced, and they will be finished in the next few decades. So, it is essential to choose alternative energy sources that will be renewable, clean, abundant, and refilled in the human cycle. Solar energy, wind energy, biomass, tidal energy, etc., are examples of such kinds of energy sources. Among them, solar energy is the most acceptable renewable energy source due to the geographical position of the sun. The sun is the main source of all energy on earth. The light from the sun takes only 8.5 min to reach on the Earth's surface though it is far away from the Earth which is about 90 miles. There are so many factors like latitude of the location, time of the day, whether of the day, etc. The solar energy incident of a location depends. Airmass (AM) is the path length which light takes through the atmosphere normalized to the shortest possible path length. In a solar cell study, AM 1.5 is the used solar spectrum. For non-terrestrial applications, AM0 is used.

The literature says 71% of the solar energy which reached the Earth's surface is absorbed by the ocean, land surface, and atmosphere [1, 2]. The main challenge is to convert solar energy into a usable form. A solar cell is a device that acts upon light fallen on it. Electrons, the majority carrier in n-type, are combined with holes, and holes are the majority carrier in p-type, leaving positive ions in n-type and negative ions on the p-side. This charge near the junction forms an electric field that restricts further recombination of electrons and holes mostly in the depletion and the quasi-neutral region. When light falls with higher energy than energy between the valence and conduction bands, the electrons are forced to jump and generate an e–h pair. Electrons are now flowed toward the n-side and holes to the p-side by the influence of the field produced in the junction.

If an external circuit is connected with the solar cell, then the electrons will flow through the circuit from the n-side to the p-side, and, as a result, a current will flow from the p-side to the n-side. By making metal contacts in the solar cell, we can withdraw the power for external uses. As direct current is produced by solar cells, an inverter is connected with the solar panels to convert this current into an alternating current.

There are mainly two types of solar cells depending on the number of junctions: single-junction solar cells and multi-junction solar cells. A single-junction solar cell can absorb wavelength range from 300 to 1100 nm. The conversion efficiency of sunlight to electrical energy for a multi-junction solar cell is more significant than for a single-junction solar cell.

1.2 Formation of Silicon Wafer and Substrate-related Study

1.2.1 Refining of silicon

Silicon is the second most abundant material on the Earth after oxygen [3]. It is not majorly available in pure form in nature. In nature, silicon is available in the form of silicon dioxide (SiO_2) instead of its elemental form majorly. The manufacturing of purer silicon occurs in two stages. By removing oxygen, metallurgical grade silicon is produced, and, by further refining, semiconductor-grade silicon is produced. An intermediate grade with an impurity level between metallurgical grade and semiconductor-grade silicon is also available, which is known as solar grade silicon [4].

1.2.1.1 Metallurgical grade silicon

It is such a kind of silicon which is not fitting for electronics applications in terms of purity. For the electronics application, this kind of silicon should be additionally purified, and, thus, it can be used in solar cells and electronics industries.

A mixture of quartz and carbon is heated in an electrode arc furnace to around 1500 to 2000°C. First, silicon monoxide (SiO) and carbon monoxide (CO) are produced in gaseous form at the bottom of the furnace, where the temperature is around 1900°C.

$$SiO_2 + C \rightarrow SiO + CO.$$

Then, silicon monoxide (SiO) and carbon monoxide (CO) flow to the cooler part of the furnace, which produces silicon carbide (SiC).

$$SiO + CO \rightarrow SiC + CO.$$

The reaction between SiO_2 and SiC can produce this type of silicon.

Finally, through a tap hole, the liquefied silicon is evacuated to the casting area, and then it will be stiffened with a purity of roughly 99%. The major impurities are used in the process, such as aluminum (Al) and iron (Fe) [5].

1.2.1.2 Semiconductor-grade silicon

Semiconductor-grade silicon (SGS) or electronic-grade silicon (EGS) is a highly purified version of MGS with extremely low impurities. The formation of EGS from MGS is expensive. The common way to produce EGS is to transform MGS, which can be refined by purification and then decomposed to reform maximum purity silicon. Trichlorosilane ($SiHCl_3$) is mostly used as an intermediate compound for EGS formation. It can be handled effortlessly and also can be stored in solid form.

In a fluidized bed reactor, powdered MG-Si reacts at around 300°C to make trichlorosilane.

$$\text{Si (solid)} + \text{HCl (gas)} \rightarrow SiHCl_3 + H_2.$$

During the reaction, the major impurities present in MGS, which are iron (Fe), aluminum (Al), and boron (B), also react.

$$SiHCl_3 + H_2 \rightarrow Si + HCl.$$

This process is first developed by Siemens; so it is often known as Siemens process [6]. In this process, the co-product, rather than Si, is HCl which reacts with $SiHCl_3$ and produces silicon tetrachloride ($SiCl_4$). This is the disadvantage of this Siemens process which is the poor efficiency of silicon and chlorine consumption.

The produced SGS are broken into pieces that are used as feedstock for the crystallization process. For solar cell fabrication, alternative processes are investigated to form upgraded metallurgical grade silicon or solar grade silicon.

1.2.2 Crystal growth

1.2.2.1 Single crystal

The grain size of a single-crystalline solar cell is > 10 cm. Single crystalline substrates are basically of two types which are defined by the process by which they are made. There are two types of silicon wafers: Czochralski (Cz) and float-zone (Fz) silicon wafers.

a) Czochralski (Cz) Silicon

This method is widely used for growing semiconductor silicon. High purity SG-Si is liquefied in a quartz crucible at the temperature of 1425°C. For making the semiconductor p-type or n-type, the impurity atom boron (B) or phosphorus (P) is added to molten silicon in a precise amount. A rod-shaped seed crystal is dipped in the silicon, and the rod is pulled upward slowly and rotated simultaneously [7]. A large cylindrical single crystal ingot can be produced by keeping the perfect temperature control and rate of pulling the rod. This process is performed in an inert chamber (e.g., quartz) and an inert atmosphere (e.g., argon). The main benefit of this method is that it can produce a large single crystal (ingot up to length of 2m); so it has wide use in the semiconductor industry. There is no direct contact between the crystal and the crucible wall for which an unstressed single crystal can be produced Figure 1.1 shows a typical schematic of Czochralski pullar.

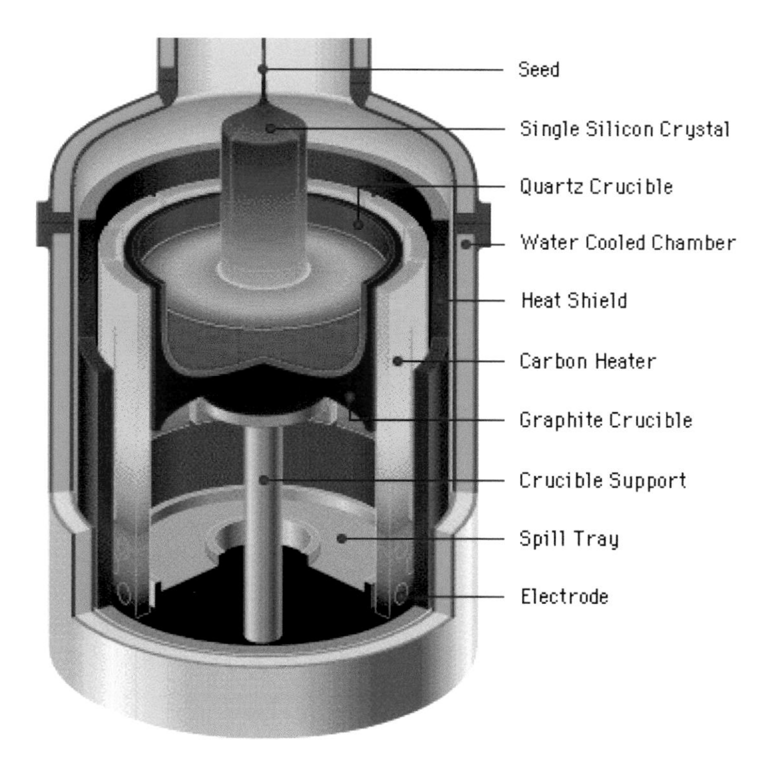

Figure 1.1 Schematic representation of Czochralski crystal puller.

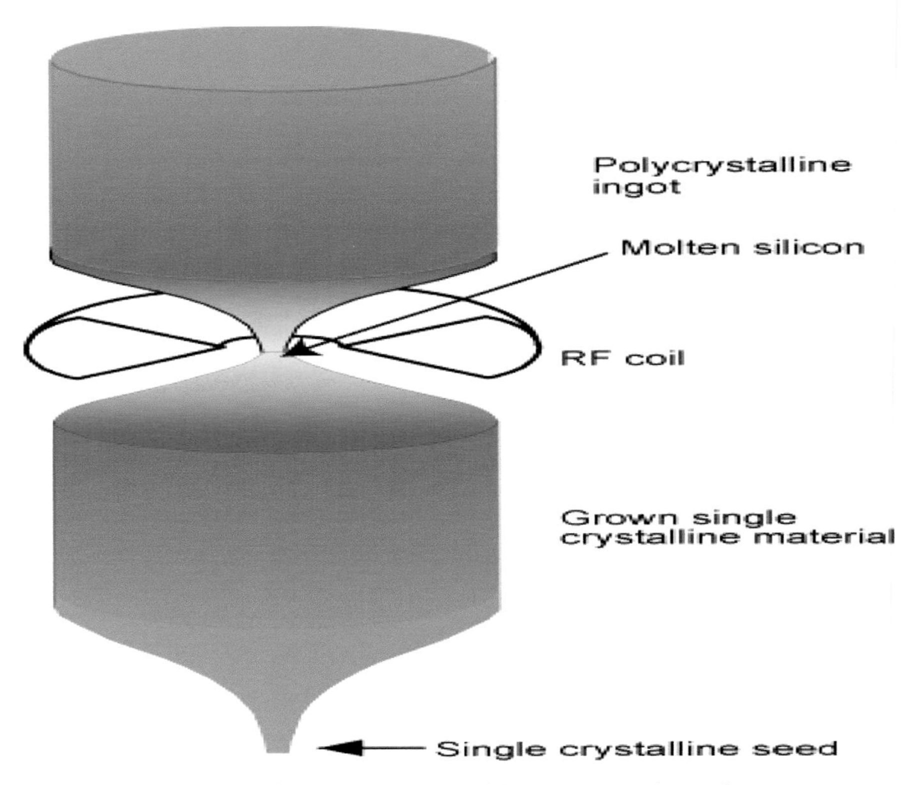

Figure 1.2 Schematic setup of float-zone crystal growth.

b) Float-Zone (Fz) Silicon

Float-zone (Fz) silicon is more pure crystal than Cz-type silicon which is obtained by zone melting [9]. Cz wafers contain a large amount of oxygen (O_2) which reduces the minority carrier lifetime and, as a result, reduces the open-circuit voltage, current, and efficiency in the solar cell. Along with oxygen, other impurities can form the oxides, which may be active at higher temperatures and make the wafers sensitive at a higher temperature. Fz silicon overcomes all these problems [8]. The size of a single crystal produced by the Fz process is smaller than the Cz process. Fz wafers are mainly used for laboratory purposes rather than industrial purposes. The cost of the Fz wafer is more than the Cz wafer. Figure 1.2 shows the schematic setup of float-zone crystal growth.

1.2.2.2 Multi-crystalline silicon

Multi-crystalline silicon (mc-Si) is a kind of silicon material with multiple grains of crystals with multiple orientations, shapes, and grain sizes larger than 1 mm. This type of silicon production is cheaper in terms of fabrication than the Cz process. It is produced by directional solidification in a quartz crucible. As the grain boundaries and the metallic impurities present in this kind of silicon, the conversion efficiency is lower than the monocrystalline silicon. The grain boundaries can also reduce the solar cell performance by creating multiple shunt paths.

1.2.3 Wafer slicing and polishing method

Ingots are sliced into smaller blocks, followed by pieces into wafers. After cutting by saw, small cracks may be seen in the wafers, which are around 10-μm deep. These cracks may increase the surface recombination and decrease the mechanical strength of the wafer [10, 11]. These saw multiple chemical baths now remove damage in order to make the wafer strong and flatter, which is known as a chemo-mechanical treatment. The edges are now corrected using an edge-grinding procedure which reduces the breakage probability.

Now wafers are treated to get uniform the elimination of saw damage, surface defects, and stress accumulated during the cutting process. After that, these wafers go through a series of chemical processes for cleaning. The last step for cleaning and polishing is done in a clean room where the wafers are cleaned using ultra-pure water and chemicals to produce a mirror-like surface that removes all the micro-cracks, saw damages, surface topography scratches, etc.

1.3 Cell Processing Technologies

After polishing, the wafers are now ready for cell processing. Following are the processing steps for producing basic solar cells.

1.3.1 Texturing

After cleaning, the front surface of the wafers is now shiny and reflects maximum light falling on the surface. A good light-trapping scheme must be incorporated to avoid maximum reflection from the front surface. Texturing the wafer surface is one of the front-side light trapping schemes to minimize the reflection of light from the surface, and by multiple reflections on the surface, it enhances the light absorption in the wafers by entering maximum light in the wafer and hence increasing the optical path length. The solar cell mainly absorbs the maximum part of the visible and ultra-violet region of the solar spectrum. Surface texturization can help to absorb a part of

Figure 1.3 Texturization unit available at IIEST Shibpur.

the longer wavelength. Thus, by texturing, the surface optical loss can be minimized.

Texturization is done either by chemical or physical methods in laboratories or industries [12]. Figure 1.3 shows the laboratory setup of the texturization unit (available at IIEST Shibpur). Isotropic wet-etching, either by acid or alkaline solution, is the main idea behind the chemical method [13]. It is found that monocrystalline wafers can be textured by a weak solution of 2% NaOH or KOH with 7% IPA at 80–85°C, which results in distributed pyramids in different etch rates [12, 14, 15]. After texturization, three cleaning processes are there: RCA (Radio Corporation of America) I, RCA II, and Piranha. In RCA I, the wafers are cleaned in the solution of ammonium hydroxide (NH_4OH), hydrogen peroxide (H_2O_2), and water (H_2O) in the ratio of 1:1:6 at 80–85°C for 15 min. After that, the wafers are dipped in 1% HF and then rinsed with DI water. In RCA II, the wafers are cleaned in the solution of HCl and H_2O_2. The HCl and H_2O_2 at a ratio of 1:1:5 are used to treat the surface at 80-85°C for 15 min. The wafers are again dipped in 1% HF solution and rinsed with DI water. They sometimes depend upon the wafer quality; a third cleaning process is required, which is known as Piranha. In Piranha, the wafers are dipped in the solution of H_2SO_4 and H_2O_2 at a ratio of 5:1 for more or less than 20 min. Texturing can also be done by physical methods.

1.3.2 Emitter diffusion process

Junction formation is a very important step in solar cell fabrication. Junction formation can be done in three methods in the solar cell: diffusion, ion

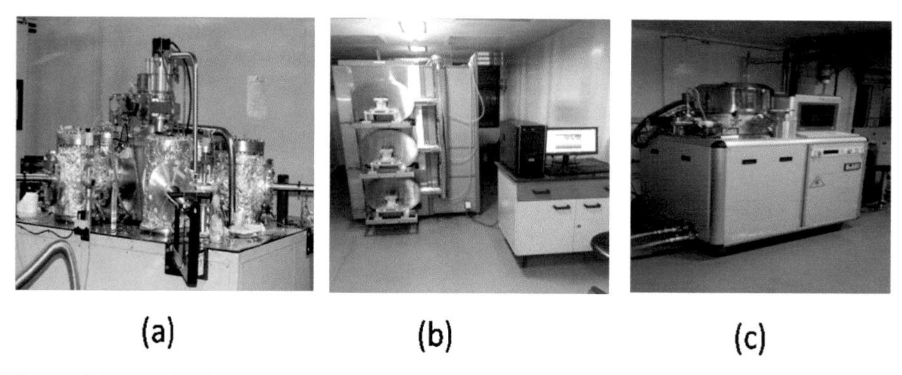

Figure 1.4. (a) PECVD unit. (b) Thermal oxidation diffusion furnace. (c) DC and RF sputtering unit (available at IIEST Shibpur).

implantation, and doping using alloy. The third way of diffusion is hardly used in today's industry.

Diffusion is the most important process to form the junction. In the diffusion process, the atoms from higher concentrations move slowly to lower concentrations at random motion. In diffusion, the surface of the silicon substrate is exposed to high concentrated dopant, and, thus, by solid-state diffusion, the dopant moves into the substrate. In a p-type wafer, phosphorus (P) is used as a dopant to form a junction, and in an n-type wafer, boron (B) is used as a dopant.

Temperature and the gaseous environment are two major factors that can influence the diffusion process [19]. The n+ emitter layer is formed by phosphorus doping at the front surface. Figure 1.4 shows the laboratory setup of (a) PECVD unit, (b) thermal oxidation diffusion furnace unit, and (c) DC and RF sputtering unit. First, in the pre-deposition stage, phosphorus is taken in $POCl_3$ form at 800°C for approximately 20 min. Nitrogen (N_2) and oxygen (O_2) are taken as carrier gas. The presence of oxygen forms SiO_2 on the silicon wafer surface. $POCl_3$ reacts with oxygen and forms P_2O_5. The P_2O_5 formed reacts with SiO_2 and forms PSG.

$$POCl_3 + O_2 \rightarrow P_2O_5 + Cl_2.$$

In the drive-in stage, the wafers are heated in the presence of nitrogen (N_2) at 830°C for 30 min, which forms the junction by penetrating the phosphorous atom in the n+ layer.

$$P_2O_5 + Si \rightarrow P + SiO_2$$

$$P + Si \rightarrow \text{n-type doped Si.}$$

Ion implantation is another junction formation technique which slowly replaces the diffusion technique in today's world [23, 24]. The ion implantation technique is the most accurate method for junction formation. In this method, the selected impurity ions are added to the semiconductor. The impurity ions are stimulated to high energy levels to form the ion implantation into the semiconductor (using an electric and magnetic field). In this method, the impurity ions are injected deep inside the semiconductor, whereas in the case of diffusion, the impurity ions are placed only at the surface.

1.3.3 PSG removal and edge isolation

After diffusion, PSG is formed at the surface. This phosphosilicate glass (PSG) can be removed by dipping the wafers in a 5% HF solution for 2 min, followed by a rinse with DI water.

The edge isolation is needed to remove the phosphorus formed around the edge of the cell so that the front emitter is electrically isolated from the rear side. Edge isolation can be done by reactive ion etching [25, 26] or wet etching [27, 28], followed by rinse with DI water. Figure 1.5 shows the laboratory setup of the reactive ion etching (RIE) unit.

Figure 1.5 Reactive ion etching unit.

1.3.4 Anti-reflection coating (ARC) to reduce the reflection loss

The name anti-reflection coating suggests that it can reduce the reflection of light from the front surface (and front and back for bifacial solar cells). ARC can cancel the reflective waves by destructive interference. If two adjacent materials have a large discrepancy in refractive indices, then the reflection will be large at the surface. It is applied on the surface to reduce reflectivity and enhance photon absorption in a certain wavelength range [29–36].

Silicon nitride (SiN_x) is the most usable ARC for industrially available solar cells. It is also desirable to reduce the front surface defects (and minority carrier recombination). The hydrogenated SiN_x works at a high-quality surface passivator [37]. ARC is typically deposited by sputtering or chemical vapor deposition (CVD) and plasma-enhanced chemical vapor deposition (PECVD) process. In this process, SiH_4 and NH_3 are the precursor gasses.

$$SiH_4 + NH_3 \rightarrow Si_3N_4 + H_2.$$

Most of the hydrogen formed in the reaction evaporates. A small part of this hydrogen diffuses in the bulk and attaches to the bulk defects, which can improve the carrier lifetime [40–43]. So, deposition of SiN_x reduces reflection and increases bulk passivation [38–44].

Previously, titanium dioxide (TiO_2) was used as a good ARC coating, but it cannot provide surface or bulk passivation.

1.3.5 Metallization and screen printing

In solar cell manufacturing, metallization is the second most expensive process (as we use Ag and Al paste, but Cu can also be the alternative choice). Metallization is done by the screen printing method followed by co-firing. Screen printing is a simple, inexpensive, cost-effective, and robust metallization method. Contacts are made on the surface, which can collect the charge carriers. The printing parameters are the paste composition, paste viscosity, emulsion thickness, and paste temperature [45–51]. Figure 1.6 shows the laboratory setup of (a) screen printing and (b) belt furnace unit.

First, the cell is placed for front-side metallization, and a particular screen pattern is printed using silver (Ag) paste. The unnecessary organic components in the paste are removed in the belt furnace to keep away from all the markings. The cells are again loaded in a conveyer belt for back screen printing. For back-side metallization, aluminum is a commonly used metal.

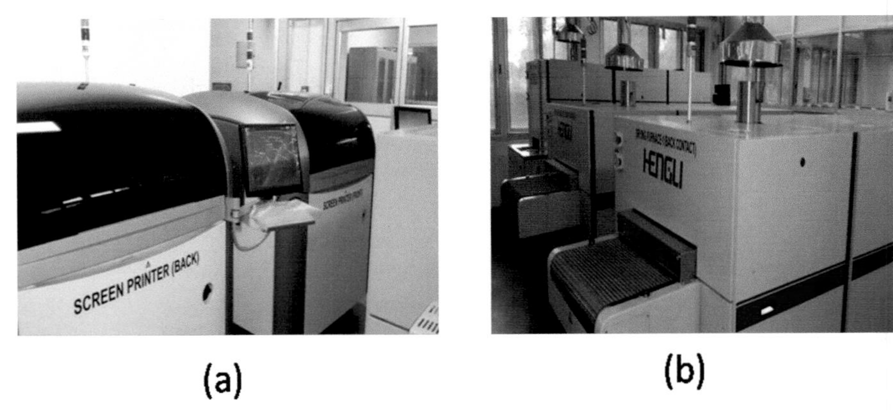

Figure 1.6 (a) Screen printing unit and (b) belt furnace unit (at IIEST Shibpur).

By metallization, contacts are made. The electrical connection can be made by front contacts which are fingers and busbars. In the rear side, the aluminum creates back surface fields.

1.3.6 Testing

This is the last step of solar cell manufacturing. It includes the determination of cell parameters using a solar power meter which is described in Figure 1.7 of the individual solar cell. The solar cells are connected in series to form the solar module. Along with avoiding hot-spot heating, the reverse breakthrough characteristics are also determined. Solar cells are finally sorted to minimize the possible mismatch losses in a module [52–54].

Solar modules are checked finally before packaging. The modules are tested by measuring the *I–V* characteristics at room temperature (25°C) into 1 sun radiation at AM1.5 radiation spectra (which is known as standard test condition (STC)).

1.4 Conclusion

This chapter explains the different steps involved in solar cell fabrication in detail. Metallurgical grade silicon is mainly used in solar cell fabrication. The solar cell cost is majorly dependent upon the material cost; so to minimize the solar cell cost, it is important to reduce the wafer thickness in order to reduce the material consumption. If the wafer thickness gets reduced, then the proper light management scheme must be incorporated to absorb the inactive photon region of the solar cell to maintain cell efficiency. The optimized front grid design can minimize the total power losses

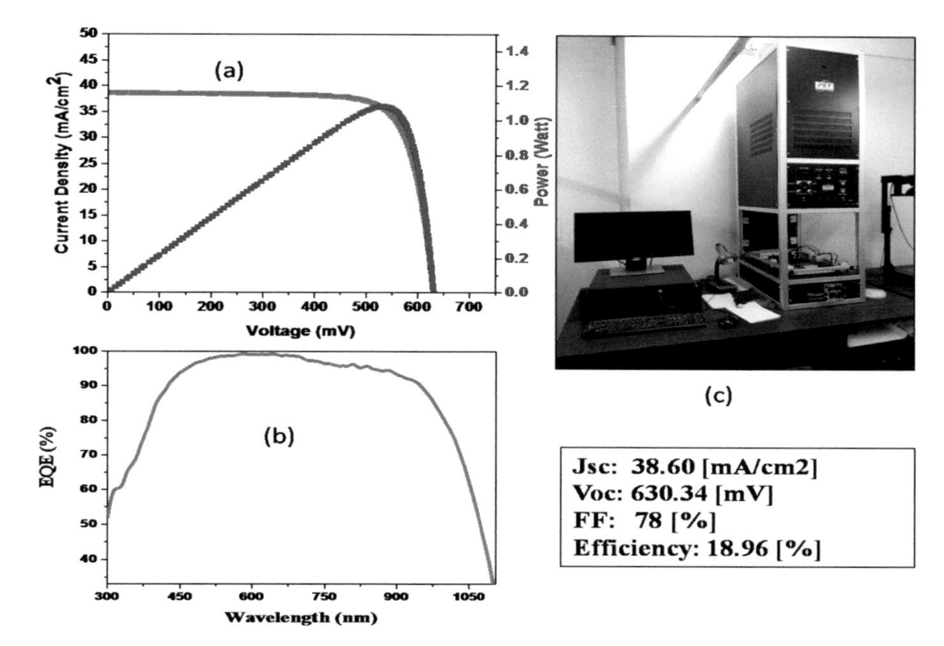

Figure 1.7 (a) *I-V* characteristics measured from the solar simulator. (b) EQE measurement. (c) Solar power meter.

(electrical power losses and shading losses); hence, the efficiency can be increased.

References

[1] K. E. Trenberth, J. T. Fasullo and J. Kiehl, Earth's global energy budget, Bull. Am. Meteorol. Soc., 2009, Vol.90, Issue 3 311.

[2] R. Lindsey, Climate and Earth's Energy Budget, United States Geological Services (USGS) Fact Sheet, Image retrieved from https://Earthobservatory.nasa.gov/Features/EnergyBalance/printall.php.

[3] R. M. Izatt, S. R. Izatt, R. L. Bruening, N. E. Izatt and B. A. Moy, Challenges to achievement of metal sustainability in our high-tech society, Chem. Soc. Rev., 2014, 43, 2451.

[4] S. Pizzini, Towards solar grade silicon: Challenges and benefits for low cost photovoltaics, Solar Energy Mater. Solar Cells, 2010, 94(9), 1528–1533

[5] C. B. Honsberg and S. G. Bowden, Refining silicon, 2019, page on www.pveducation.org

[6] Y. Dazhou, Siemens process. In: D. Yang (eds.) Handbook of Photovoltaic Silicon. Springer, Berlin, Heidelberg, 2018, https://doi.org/10.1007/978-3-662-52735-1_4-1

[7] Wikipedia Contributors. (2020, May 22). Czochralski method. In: Wikipedia, The Free Encyclopedia. Retrieved 22:36, October 2, 2020, from https://en.wikipedia.org/w/index.php?title=Czochralski_method&oldid=958243367.
[8] Wikipedia Contributors. (2020, February 12). Float-zone silicon. In: Wikipedia, The Free Encyclopedia. Retrieved 22:41, October 2, 2020, from https://en.wikipedia.org/w/index.php?title=Float-zone_silicon&oldid=940355684
[9] G. Fisher, M. R. Seacrist and R. W. Standle, Silicon crystal growth and wafer technologies, Proc. IEEE, 2012, 100, 1454.
[10] D. H. Neuhaus and A. Munzer, Industrial Silicon Wafer Solar Cells, Adv. Optoelectron., 2007, Volume 2007 24521.
[11] D. L. King and M. E. Buck, Experimental optimization of an anisotropic etching process for random texturization of silicon solar cells, Proceedings of the 22nd IEEE Photovoltaic Specialists Conference (PVSC '91), Las Vegas, NV, USA, Oct. 1991, vol. 1, p. 303.
[12] M. F. Abdullah, M. A. Alghoul, H. Naser, N. Asim, S. Ahmadi, B. Yatim and K. Sopian, Research and development efforts on texturization to reduce the optical losses at front surface of silicon solar cell, Renewable Sustainable Energy Rev., 2016, 66, 380.
[13] E. Vazsonyi, K. de Clercq, R. Einhaus, E. van Kerschaver, K. Said, J. Poortmans, J. Szlufcik and J. Nijs, Improved anisotropic etching process for industrial texturing of silicon solar cells, Solar Energy Mater. Solar Cells, 1999, 57, 179-188.
[14] H. Saha, S. K. Datta, K. Mukhopadhyay and S. Banerjee, Influence of surface texturization on the light trapping and spectral response of silicon solar cells, IEEE Trans. Electron Devices, 1992, 39, 5.
[15] U. Gangopadhyay, K. Kim, A. Kandol, J. Yi and H. Saha, Role of hydrazine monohydrate during texturization of large-area crystalline silicon solar cell fabrication, Sol. Energy Mater. Sol. Cells, 2006, 90, 3094.
[16] U.S. Geological Survey, Mineral Commodity Summaries, January 2017.
[17] B. Gao, S. Nakano and K. Kakimoto, Influence of reaction between silica crucible and graphite susceptor on impurities of multi-crystalline silicon in a unidirectional solidification furnace, J. Cryst. Growth, 2011, 314, 239.
[18] A. F. B. Braga, S. P. Moreira, P. R. Zampieri, J. M. G. Bacchin and P. R. Mei, New processes for the production of solar-grade polycrystalline silicon: A review. Sol. Energy Mater. Sol. Cells, 2008, 92, 418.
[19] G. Masetti, S. Solmi and G. Soncini, On phosphorus diffusion in silicon under oxidizing atmospheres, Solid-State Electronics, 1973, 16, 419-421.

[20] D. Kumar, S. Saravanan and P. Suratkar, Effect of oxygen ambient during phosphorous diffusion on silicon solar cell, J. Renewable Sustainable Energy, 2012, 4(8), 033105-033105.
[21] H. Uchida, Y. Ieki, M. Ichimura and E. Arai, Retarded diffusion of phosphorus in silicon-on-insulator structures, Japanese J. Appl. Phys., 2000, 39, L137-L140. Fabrication of Crystalline Silicon Solar Cell with Emitter Diffusion, SiNx Surface Passivation and Screen Printing of Electrode. http://dx.doi.org/10.5772/51065 129.
[22] M. Popadic, L. K. Nanver and T. L. M. Scholtes, Ultra-shallow dopant diffusion from pre-deposited RPCVD monolayers of arsenic and phosphorus, Proceedings of the 15th International Conference on Advanced Thermal Processing of Semiconductors, Oct. 2-5, 2007, pp. 95–100.
[23] H. Hieslmair, I. Latchford, L. Mandrell, M. Chun and B. Adibi, Ion Implantation for Silicon Solar Cells. Intevac, Santa Clara, CA, USA, 2012, http://www.pv-tech. org.
[24] M. Fischer, A. Metz and S. Raithel, Challenges in c-Si Technology for Suppliers and Manufacturers, Proceedings of the 27th European Photovoltaic Solar Energy Conference (EU PVSEC), Frankfurt, Germany, 2012, 527–532.
[25] J. Arumughan, T. Pernau, A. Hauser and I. Melnyk, Simplified edge isolation of buried contact solar cells, Solar Energy Mater. Solar Cells, 2005, 87, 705–714.
[26] H. Jansen, H. Gardeniers, M. de Boer, M. Elwenspoek and J. Fluitman, A survey on the reactive ion etching of silicon in microtechnology, J. Micromechanics Microengineering, 1996, 6, 14-28.
[27] K. Yamamura and T. Mitani, Etching characteristics of local wet etching of silicon in HF/HNO3 mixtures, Surface Interface Analysis, 2008, 40, 011–013.
[28] M. Steinert, J. Acker, M. Krause, S. Oswald and K. Wetzig, Reactive species generated during wet chemical etching of silicon in HF/HNO3 mixtures, J. Phys. Chem. B, 2006, 110, 11377–11382.
[29] B. S. Richards, Comparison of TiO2 and other dielectric coatings for buried-contact solar cells: a review, Prog. Photovoltaics Res. Appl., 2004, 12, 253.
[30] R. Narzary, S. Maity and P. Pratim Sahu, Coupled ZnO–SnO2 nanocomposite for efficiency enhancement of ZnO–SnO2/p-Si heterojunction solar cell, IEEE Trans. Electron Devices, 2021, 68(2), 610–617.
[31] S. Maity, B. Das, R. Maity, N. Pratap Maity, K. Guha and K. Srinivasa Rao, Improvement of quantum and power conversion efficiency through electron transport layer modification of ZnO/perovskite/PEDOT: PSS

based organic heterojunction solar cell, Solar Energy J., 2019, 185, 439–444.
[32] S. Maity and T. Thomas, Hole collecting treated graphene layer and PTB7:PC71BM based bulk-heterojunction OPV with improved carrier collection and photovoltaic efficiency, IEEE Trans. Electron Devices (Accepted), 2018, DOI:10.1109/TED.2018.2864537
[33] D. Muchahary and S. Maity, High-efficiency thin film ZnMgO/ZnO solar cell simulation approach: Temperature dependency, BSF and efficient small signal analysis, Superlattices Microstruct., 2017, Volume 109, Pages 209–216.
[34] S. Maity, C. Tilak Bhunia and P. Pratim Sahu, Improvement in optical and structural properties of ZnO thin film through hexagonal nanopillar formation to improve the efficiency of Si-ZnO heterojunction solar cell, J. Phys. D Appl. Phys., 49(20), 1–9.
[35] S. Maity and S. Sahare, Silicon nitride deposition using inductive coupled plasma chemical vapor deposition technique to study optoelectronics behavior for solar cell application, Optik - Int. J. Light Electron Optics, Elsevier, 2016, 127(13), 5240–5244.
[36] N. S. Beattie, P. See, G. Zoppi, P. M. Ushasree, M. Duchamp, I. Farrer, D. A. Ritchie and S. Tomic, Quantum Engineering of InAs/GaAs Quantum Dot Based Intermediate Band Solar Cells, ACS Photonics, 2017, 4, 2745.
[37] H. Mackel and R. Ludemann, Detailed study of the composition of hydrogenated SiNx layers for high-quality silicon surface passivation, J. Appl. Phys., 2002, 92, 602–609.
[38] G. Aberle, Overview on SiN surface passivation of crystalline silicon solar cells, Sol. Energy Mater. Sol., 2001, 65, 239.
[39] W. Soppe, H. Rieffe and A. Weeber, Bulk and surface passivation of silicon solar cells accomplished by silicon nitride deposited on industrial scale by microwave PECVD, Prog. Photovoltaics, 2005, 13, 551.
[40] Z. Chen, A. Rohatgi, R. O. Bell and J. P. Kalejs, Defect passivation in multi-crystalline-Si materials by plasma-enhanced chemical vapor deposition of SiO2/SiN coatings, Appl. Phys. Lett., 1994, 65, 2078.
[41] J. Hong, W. M. M. Kessels, W. J. Soppe, A. W. Weeber, W. M. Arnoldbik and M. C. M. Van De Sanden, Influence of the high-temperature "firing" step on high-rate plasma deposited silicon nitride films used as bulk passivating antireflection coatings on silicon solar cells, J. Vac. Sci. Technol., B: Microelectron. Nanometer Struct., 2003, 21, 2123.
[42] K. Kimura, Recent developments in polycrystalline silicon solar cell, Proceedings of the 1st International Photovoltaic Science and Engineering Conference, Kobe, Japan, 1987, p. 37.

[43] K. V. Ravi, R. C. Gonsiorawski and A. R. Chaudhuri, Progress in EFG Technology for Low Cost Photovoltaics, Proceedings of the 7th IEEE Photovoltaic Specialists Conference, Las Vegas, NV, USA, 1985, p. 1222.
[44] F. Duerickx and J. Szlufcik, Defect passivation of industrial multi-crystalline solar cells based on PECVD silicon nitride, Sol. Energy Mater. Sol., 2002, 72, 231.
[45] P. J. Holmes and R. G. Loasby, Handbook of Thick Film Technology, Electrochemical Publications, Glasgow, Scotland, UK, 1976.
[46] G. Schubert, F. Huster and P. Fath, Current transport mechanism in printed Ag thick film contacts to an n-type emitter of a crystalline silicon solar cell, Proceedings of the 19th European Photovoltaic Solar Energy Conference (EU PVSEC '04), Paris, France, 2004, p. 813.
[47] A. Huthig, Hybridintegration, In: H. Reichl (ed.). Heidelberg, Germany, 1986.
[48] C. Ballif, D. M. Huljić, A. Hessler-Wysser and G. Willeke, Proceedings of the 29th IEEE Photovoltaic Specialists Conference (PVSC '02), New Orleans, LA, USA, 2002, p. 360.
[49] C. Ballif, D. M. Huljić, G. Willeke and A. Hessler-Wysser, Silver thick-film contacts on highly doped n-type silicon emitters: Structural and electronic properties of the interface, Appl. Phys. Lett., 2003, 82, 1878.
[50] G. Schubert, B. Fischer and P. Fath, Formation and nature of Ag thick film front contacts on crystalline silicon solar cells, Proceedings of Photovoltaics in Europe Conference (PV '02), Rome, Italy, 2002, p. 343.
[51] G. Schubert, F. Huster and P. Fath, Current transport mechanism in printed Ag thick film contacts to an n-type emitter of a crystalline silicon solar cell, Proceedings of the 14th International Photovoltaic Science and Engineering Conference (PVSEC '04), Bangkok, Thailand, Jan. 2004, p. 441.
[52] D.-H. Neuhaus, R. Mehnert and G. Erfurt, Loss analysis of solar modules by comparison of IV measurements and prediction from IV curves of individual solar cells, Proceedings of the 20th European Photovoltaic Solar Energy Conference, Barcelona, Spain, 2005, p. 1947.
[53] D.-H. Neuhaus, J. Kirchner and R. Mehner, Impact of shunted solar cells on the IV characteristics of solar modules, Proceedings of the 21st European Photovoltaic Solar Energy Conference, Dresden, Germany, 2006, p. 2556.
[54] D.-H. Neuhaus, F. Dreckschmidt and R. Ludemann, Suitability tests for solar cell measurment in the final quality check of an industrial production process, Proceedings of the 19th European Photovoltaic Solar Energy Conference, Paris, France, 2004, p. 817.

Chapter 2

Uncertainty-Based Battery Sizing in District Energy Community with Distributed Renewable Systems

Yuekuan Zhou[1,2]

[1]Sustainable Energy and Environment Thrust, Function Hub, The Hong Kong University of Science and Technology, China
[2]Department of Mechanical and Aerospace Engineering, The Hong Kong University of Science and Technology, Clear Water Bay, China
Corresponding author: yuekuanzhou@ust.hk; yuekuan.zhou@connect.polyu.hk

Abstract

Along with fossil fuel depletion and deteriorated environmental conditions, renewable energy has attracted widespread interest in cleaner power production and renewable and sustainable transition. Unlike traditional power plants with easy controllability and demand-based design principles, renewable power fluctuates and is intermittent, such as solar-dependent PV systems, wind turbines, etc. Therefore, challenges are imposed on flexible strategies to enhance renewable self-consumption, demand coverage, and system self-independence and reduce grid interaction. This chapter formulated a district energy community for renewable and sustainable transitions. Using the component-wise Metropolis–Hastings algorithm, a two-dimensional Markov chain Monte Carlo (MCMC) multivariate distribution was adopted for scenario uncertainty quantification with the bivariate distribution. Through feature extraction and classification, data-driven models were structured and then trained by the cross-entropy algorithm for accurate predictions on building energy demands and renewable generations under stochastic uncertainties. Based on the dynamic difference between distributed renewable generations and building energy demands, a risk-based battery sizing approach was adopted to identify the storage capacity for each battery. The proposed approach can be applied for energy planning on battery storage systems with scenario uncertainties, providing preliminary design and operation strategies for energy management in a large-scale district energy community. Compared to the deterministic case study, the uncertainty-based approach can improve the reliability and robustness of the district energy community to improve renewable penetration and reduce grid reliance.

2.1 Introduction

Buildings' energy consumption consumes around 40% of total energy worldwide [1]. As the fundamental step in energy flexible buildings, accurate prediction of building energy consumption plays a critical role in a renewable and sustainable transition. In academia, prediction tools can be mainly classified into traditional physics-based simulation models, human-knowledge-based machine learning models, and data-driven models. Traditional physics-based simulation models are developed on simulation software, like EnergyPlus [2], TRNSYS [3], DesignBuilder [4], and so on. Due to the complexity of thousands of parameters and uncertainty for each parameter, traditional physics-based simulation models cannot accurately predict buildings' dynamic energy performance behaviors. With the rapid development of smart sensors, numerous databases can be collected and obtained from buildings, providing possibilities to integrate artificial intelligence inaccurate predictions on building energy performance. Zhou and Zheng [5] developed a data-driven model to accurately predict building energy consumption using the cross-entropy function. In addition to a single building, machine learning (ML) can also be applied in urban building energy performance prediction. Fathi *et al.* [6] applied ML for accurate energy performance prediction on an urban scale, considering space functionality share percentages and climate change. Amasyali and El-Gohary [7] comprehensively reviewed data-driven models for building energy prediction. The review presents two disadvantages of data-driven models, including the inaccuracy of prediction outside of the training range and unknown knowledge in internals and principles. In addition, a large amount of database is required for ML model training, testing, and validation. In order to address the challenge, human knowledge-based ML can decrease abundancy in data preparation and improve prediction accuracy and model interpretability to promote the applicability of ML in building performance prediction [8]. Zhang *et al.* [9] developed a hybrid deep-learning-based model with an interpretation process. Results showed that the hybrid model was more accurate than conventional models. Demand-side management (DSM) is one of the flexible strategies to improve the penetration ratio of renewable energy and promote the transition toward sustainability. The underlying mechanism is to manage energy demands in accordance with renewable generations in terms of local climate conditions, user needs, and grid requirements. DSM mainly includes thermal mass, heating, ventilation, and air-conditioning controls, smart charging on electric vehicles, and plug-in load shifting [10].

To overcome the intermittence of distributed PV systems and enhance the power supply reliability, energy storage has been widely applied in different energy forms, like sensible/latent thermal energy storage [11], electrical energy storage (electrochemical battery [12], pumped hydro [13], hydrogen [14, 15], and so on), and plug-in electric vehicles [16]. Depending on the characteristics of power generations and energy demands, the battery sizing approach is important to improve technical and social-economic performances. Yang *et al.* [17] comprehensively reviewed battery size determination approaches. The study indicated that the most suitable method for battery size determination depended on the types of renewable systems, system capacity, and applications. Note that input parameters, such as meteorological, design, and operating parameters, are full of uncertainties [18]. The deterministic scenario-based battery sizing capacity might be less useful in practice, especially considering the scenario uncertainty of each parameter. Normally, uncertainty can be classified into aleatory and epistemic uncertainty [19]. The former reflects the inherent randomness of performance behavior, and the latter refers to the fixed assumption on performance behavior estimation due to the lack of knowledge. Coppitters *et al.* [20] studied robust system configuration on a PV-battery-heat pump system with thermal storage, considering aleatory and epistemic uncertainty. The heat pump integrated system can effectively reduce epistemic uncertainty compared to the gas boiler integrated system. Note that, compared to deterministic case analysis, the stochastic uncertainty-based battery sizing will be more complicated regarding computational load and modeling complexity [21].

Studies on uncertainty-based building performance simulation are restrained in the single-building scale, whereas the diversity of energy demands from the district community level has been rarely considered in different types of buildings or the similar building type but with uncertainty in thermo-physical (such as thermal insulation of building envelopes) and operating parameters (such as supply and return water temperature in cooling/heating systems). Compared to the single building performance prediction, the community-level based performance prediction is much more complicated in both modeling development and simulation processes. The machine learning method can simplify the sophisticated nonlinear building performance prediction regarding different types of buildings and scenario

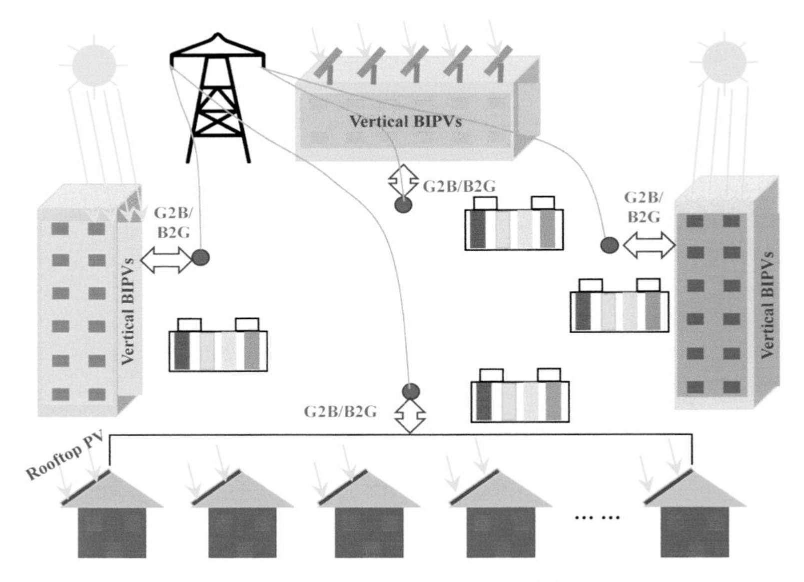

Figure 2.1 System configuration of a district community with distributed renewable systems and battery storage.

uncertainties. Second, uncertainty-based prediction on distributed renewable systems has been rarely studied on a district community scale. Third, studies on battery sizing fail to consider the uncertainty of input parameters, whereas the parameter uncertainty is critical and necessary to be considered in the battery sizing process.

Contributions of this study include 2.1) the development of a machine-learning-based surrogate model for uncertainty-based building energy predictions in large-scale district communities with high efficiency and accuracy; 2.2) uncertainty-based power production of distributed PV systems with a machine-learning-based surrogate model; 2.3) uncertainty-driven battery sizing approach for design and planning on district energy community.

2.2 Methodology

Figure 2.1 demonstrates the system configuration of a district community, consisting of distributed renewable systems (rooftop PVs and building integrated PVs), battery storage, and a micro-grid. Considering the intermittent solar power generation and stochastic demand in each building, electrochemical batteries are designed to improve the renewable penetration and enhance the demand coverage of the district community. Furthermore, the function of the micro-grid is to balance the intermittent power supply and stochastic electric load dynamically.

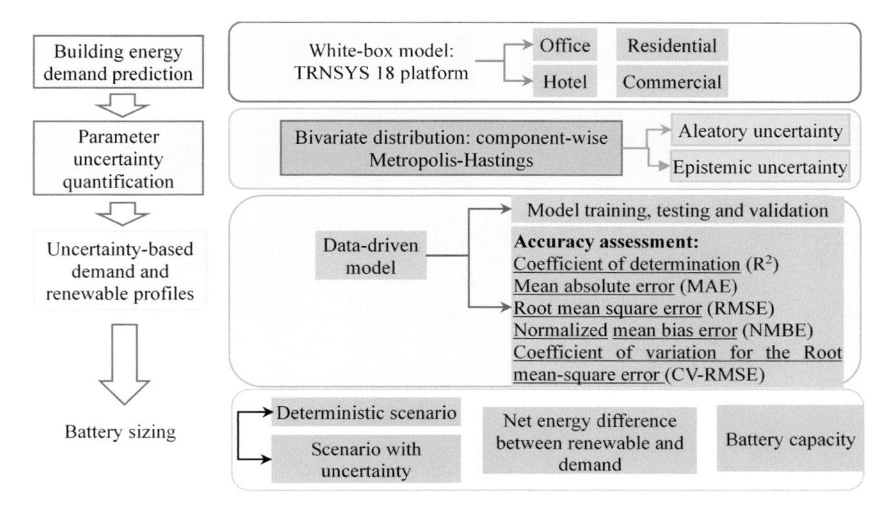

Figure 2.2 Overview of research methodology.

Figure 2.2 shows the research methodology, including energy demand prediction for different types of buildings (office, hotel, commercial, and residential buildings), parameter uncertainty quantification using the bivariate distribution method through component-wise Metropolis–Hastings, uncertainty-based demand, and renewable predictions via data-driven model, and battery sizing. Parameters for uncertainty quantification include meteorological parameters (solar radiation, ambient temperature, and wind speed), design parameters (indoor air setpoint temperature), operating parameters (supply/return water temperature), and so on. To evaluate the prediction accuracy, mathematical indicators for accuracy assessment include the coefficient of determination (R2), mean absolute error (MAE), root mean square error (RMSE), normalized mean bias error (NMBE), and coefficient of variation for the root-mean-square error (CV-RMSE). Finally, in terms of battery sizing, a comparison was conducted between deterministic and uncertainty scenarios to highlight the difference between them and illustrate the necessity for an uncertainty-based battery sizing approach.

2.3 Results and Discussions

2.3.1 Machine learning results of multi-diversified building energy demands in districts

In order to accurately predict both electrical demands and renewable generations of distributed PVs for different types of buildings in the district community, supervised machine learning was adopted to train a surrogated model,

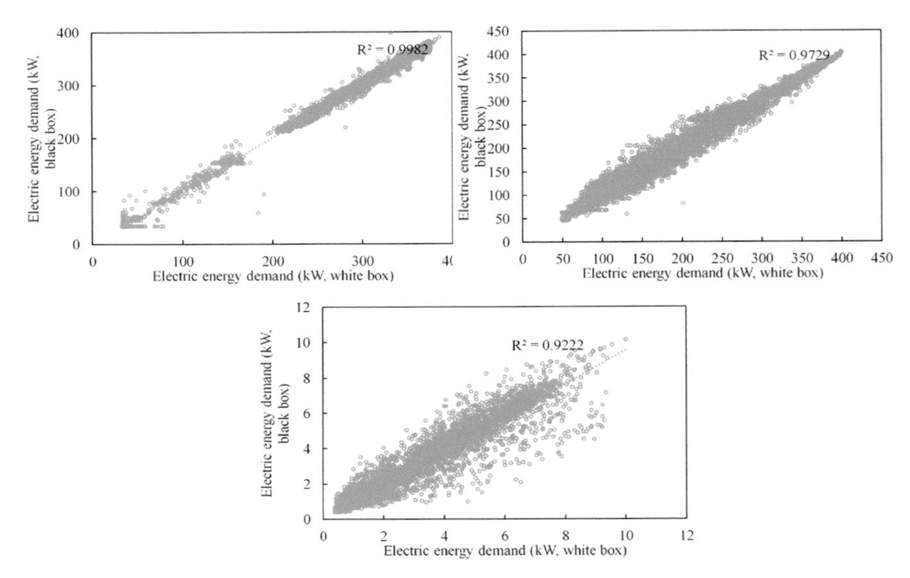

Figure 2.3 Correlation between the supervised learning predicted results and the white box model in TRNSYS in terms of building energy demands for (a) office buildings, (b) hotel buildings, and (c) residential buildings.

using the cross-entropy function, as introduced in [5]. Compared to the complex nonlinear system modeling and calculation in the mathematical model developed in TRNSYS (while boxing), the machine-learning-based surrogate model (black box) shows superiority and robustness in terms of modeling simplification, computational efficiency with no requirement on professional knowledge, such as thermodynamics, computational fluid dynamics, statistics, and so on. However, due to the lack of professional knowledge in a specific field, the black box's production accuracy will be less inferior to the white box. In this section, the comparison between black and white boxes is conducted regarding the statistic indicators.

Figure 2.3 shows the correlation between white- and black-box models regarding building energy demands for different types of buildings. As shown in Figure 2.3, the coefficients of determination are 0.9982, 0.9729, and 0.9222 for office, hotel, and residential buildings, respectively. The five statistical indicators between the black-box model and the white-box model are presented in Table 2.1. As listed in Table 2.1, the black-box model can meet the prediction accuracy requirement, as defined in ASHRAE Guideline 14 [22] (less than 10% for NMBE and less than 30% for CV-RMSE), indicating that the machine learning trained black-box model is feasible and accurate, to predict demands of multi-diversified buildings in the district community.

Table 2.1 The NMBE and CV-RMSE of the machine learning model for the demand of each building.

	R^2	MAE	RMSE	NMBE	CV-RMSE
Office building	0.9982	0.00	5.45	−0.29%	3.13%
Hotel building	0.9729	0.00	14.49	−0.77%	7.89%
Residential building	0.9222	0.00	0.58	0.97%	21.07%

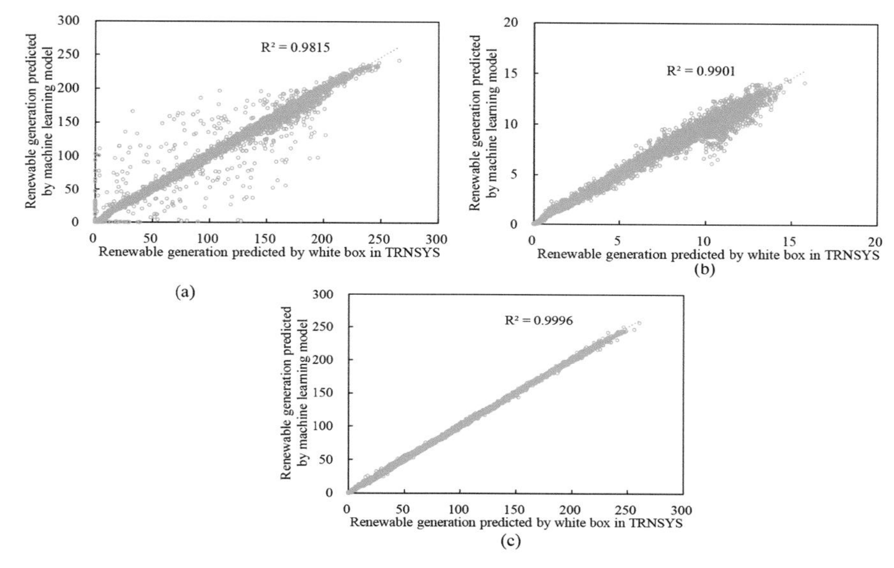

Figure 2.4 Correlation between supervised learning predicted results and the white-box model in TRNSYS in terms of renewable generations of BIPVs for (a) office and hotel buildings; (b) residential buildings; (c) commercial buildings.

Figure 2.4 shows the correlation between white and black boxes regarding renewable generations of distributed PVs in different types of buildings. As shown in Figure 2.4, the coefficients of determination are 0.9815, 0.9901, and 0.9996 for office (hotel), residential, and commercial buildings, respectively. The five statistical indicators between the black-box model and the white-box model are presented in Table 2.2. As listed in Table 2.2, the black-box model can meet the prediction accuracy requirement of the ASHRAE Guideline 14 [22] (less than 10% for NMBE and less than 30% for CV-RMSE), indicating that the machine learning trained black-box model is feasible and accurate, in terms of the prediction on renewable generations of building-integrated photovoltaics (BIPVs) of multi-diversified buildings in the district community.

Table 2.2 The NMBE and CV-RMSE of the machine learning model for renewable generation of each building.

	R^2	MAE	RMSE	NMBE	CV-RMSE
Office/hotel building	0.9815	0.00	9.48	0.067%	18.6%
Commercial building	0.9996	0.00	1.26	−0.153%	3.21%
Residential building	0.9901	0.00	0.43	1.35%	14.2%

Table 2.3 The percentage of the uncertainty of scenario parameters for electrical demand prediction.

Parameters	Percentage of uncertainty
Outdoor air temperature (T_{out})	[−13%,13%]
Solar air temperature ($T_{solar,air}$)	[−13%,13%]
Supply water temperature in the AHU cooling system ($T_{supply,AC}$)	[0.6%]
Return water temperature in the AHU cooling system ($T_{return,AC}$)	[0.6%]
Power consumption of AHU cooling chiller (PAC)	[−6%,6%]
Supply water temperature in the space cooling system ($T_{supply,SC}$)	[0.6%]
Return water temperature in the space cooling system ($T_{return,SC}$)	[0.6%]
Power consumption of space cooling chiller (PSC)	[−6%,6%]
Internal gains	[−6%,6%]
Indoor air temperature (T_{indoor})	[−6%,6%]

2.3.2 Energy demand predictions in different types of buildings with uncertainty of scenario parameters

Based on the well-trained surrogate models in Section 2.3.1, input parameters with stochastic uncertainty are prepared. The surrogate models are re-called to predict energy demands and renewable generations for different types of buildings in the district community.

2.3.2.1 Scenario Uncertainty Quantification

Considering the aleatory and epistemic uncertainties of input parameters, such as meteorological parameters, design, and operating parameters, a two-dimensional Markov chain Monte Carlo (MCMC) multivariate distribution was adopted to quantify the bivariate distribution of the component-wise Metropolis–Hastings algorithm [23]. Percentages of uncertainty for electrical demand prediction are listed in Table 2.3. As shown in Table 2.3, compared to the percentages of the uncertainty of internal gains and indoor air temperature (T_{indoor}), the outdoor air temperature and solar air temperature show a broader range. Furthermore, a positive percentage of uncertainty can be noticed in the supply/return water temperature. The reason is due to the cooling energy loss of the supply/return water. Percentages of uncertainty for

Table 2.4 The percentage of the uncertainty of scenario parameters for renewable generation of PVs.

Parameters	Percentage of uncertainty
Outdoor air temperature (T_{out})	[−13%,13%]
Total radiation on the tilted surface (I_{total})	[−13%,13%]
Beam radiation on the tilted surface (I_{beam})	[−13%,13%]
Wind speed	[−13%,13%]
The temperature of the back surface of BIPVs	[−6%,6%]

renewable generation of PVs are listed in Table 2.4. As shown in Table 2.4, compared to the exterior parameters (outdoor air temperature, solar radiation, and wind speed), the uncertainty percentages for the back surface temperature of BIPVs show a narrower range. The underlying mechanism is due to the complex impacting factors of exterior parameters, such as the shading and the heat island effect, whereas the back surface of BIPVs is less fluctuated due to the protection of the solar cell module.

2.3.2.2 Predictions on building demands and renewable generations with stochastic scenario uncertainty

Figure 2.5 shows the histogram and probability of electric demand for deterministic scenarios and stochastic uncertainty scenarios. As shown in the histogram, compared to the deterministic scenario, the stochastic scenario with uncertainty will decrease the time duration for electrical demand at a lower value and increase the time duration for electrical demand at a higher value. For instance, with respect to the office building, as shown in Figure 2.5(a) and (b), the consideration of stochastic uncertainty decreases the time duration for electrical demand within [20 kWh, 40 kWh] from 3436 to 762 hours but increases the time duration for electrical demand within [380 kWh, 400 kWh] from 11 to 650 hours. With respect to the hotel building, as shown in Figure 2.5(c) and (d), the consideration of stochastic uncertainty decreases the time duration for electrical demand within [100 kWh, 120 kWh] from 1285 to 243 hours but increases the time-duration for electrical demand within [380 kWh, 400 kWh] from 8 to 757 hours. With respect to the residential building, as shown in Figure 2.5(e) and (f), the consideration of stochastic uncertainty decreases the time duration for electrical demand within [0 kWh, 2 kWh] from 3993 to 1239 hours but increases the time duration for electrical demand within [10 kWh, 12 kWh] from 1 to 2920 hours.

Furthermore, from the probability density curve, compared to the deterministic scenario, the consideration of stochastic uncertainty will increase the average value of electric demand. The standard deviation is highly dependent

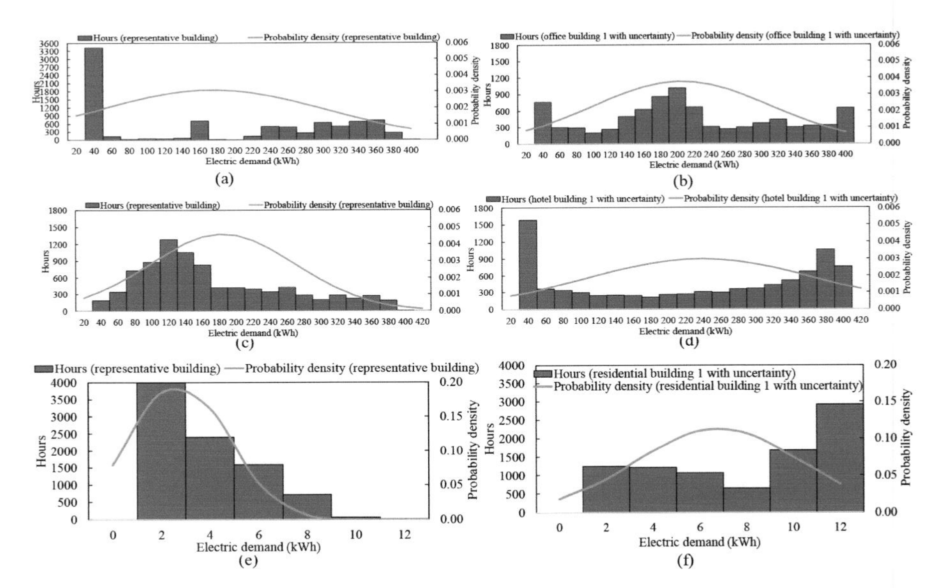

Figure 2.5 Histogram and probability of electric demand for the deterministic scenario: (a) office; (c) hotel; (e) residential building, and uncertainty scenario: (b) office; (d) hotel; (f) residential building.

on the type of building. For instance, the consideration of stochastic uncertainty increases the average electric demand from 174 to 205.1 kWh for the office building (as shown in Figure 2.5(a) and (b)) and from 183.6 to 237.9 kWh for the hotel building (as shown in Figure 2.5(c) and (d)). However, the consideration of stochastic uncertainty will decrease the standard deviation (δ) from 129.3 to 106.1 kWh for the office building (as shown in Figure 2.5(a) and (b)) but increase the standard deviation (δ) from 86.8 to 134.1 kWh for the hotel building (as shown in Figure 2.5(c) and (d)), and from 2.0 to 3.5 kWh for the residential building (as shown in Figure 2.5(e) and (f)).

Figure 2.6 shows the histogram and probability of distributed renewable generation for deterministic scenarios and stochastic scenarios with uncertainty. As shown in the histogram diagram, compared to the deterministic scenario, the stochastic scenario with uncertainty will increase the time duration for renewable generation at a lower value but decrease the time duration for electrical demand at a higher value. For instance, with respect to the office building, as shown in Figure 2.6(a) and (b), the consideration of stochastic uncertainty increases the time duration for renewable generation within [30 kWh, 60 kWh] from 615 to 666 hours but decreases the time duration for renewable generation within [210 kWh, 240 kWh] from 117 to

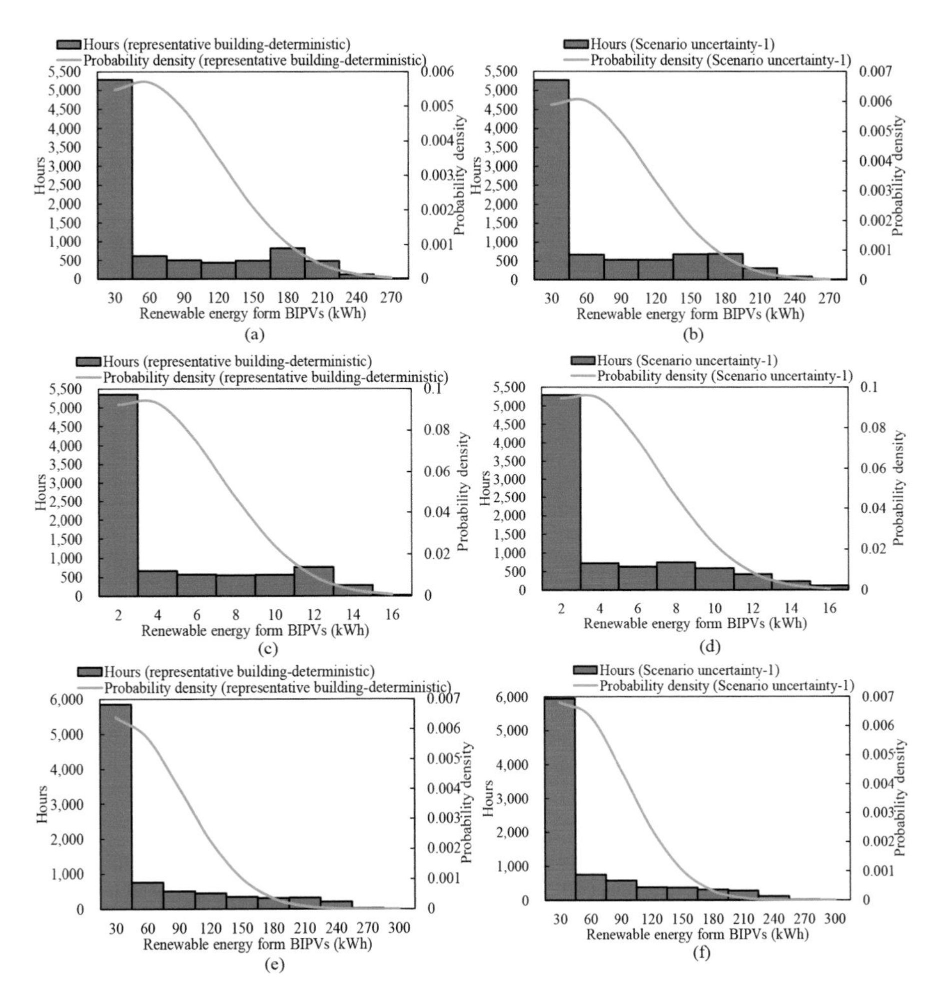

Figure 2.6 Histogram and probability of renewable generation for deterministic scenario parameters: (a) BIPVs of office and hotel buildings; (c) BIPVs of residential buildings; (e) rooftop PVs in a commercial building, and scenario parameters with uncertainty: (b) BIPVs of office and hotel buildings; (d) BIPVs of residential building; (f) rooftop PVs in the commercial building.

88 hours. With respect to the residential building, as shown in Figure 2.6(c) and (d), the consideration of stochastic uncertainty increases the time duration for renewable generation within [2 kWh, 4 kWh] from 668 to 720 hours but decreases the time duration for renewable generation within [12 kWh, 14 kWh] from 285 to 232 hours. With respect to the commercial building, as shown in Figure 2.6(e) and (f), the consideration of stochastic uncertainty increases the time duration for renewable generation within [0 kWh, 30 kWh]

from 5854 to 5944 hours but reduces the time duration for renewable generation within [210 kWh, 240 kWh] from 203 to 120 hours.

2.4 Identification of Battery Storage Capacity-Deterministic and Stochastic Uncertainty Cases

Based on the predicted energy demands and renewable generations, the battery storage capacity needs to be designed during the system planning stage. Furthermore, considering scenario uncertainty of input parameters is necessary for the battery storage capacity, as scenario parameters are full of uncertainty in realistic operation conditions. This section adopted an uncertainty-based risk acceptance methodology to size the battery storage capacity.

2.4.1 Probability density and cumulative distribution of net energy difference for deterministic and stochastic scenarios with uncertainty

Figure 2.7 demonstrates the probability density and cumulative distribution of net energy difference between renewable and demand for the deterministic and stochastic scenarios, respectively. As shown in Figure 2.7, compared to the deterministic scenario, the stochastic uncertainty will decrease the average value of the net energy difference, i.e., from −122.94 to −157.19 kWh for the office (as shown in Figure 2.7(a) and (b)), from −132.53 to −189.91 kWh for the hotel (as shown in Figure 2.7(c) and (d)), from 0.32 to −3.83 kWh for the residential building (as shown in Figure 2.7(e) and (f)), and from −27.26 to −30.11 kWh for the commercial building (as shown in Figure 2.7(g) and (h)). The underlying mechanism is that considering scenario uncertainty will decrease the renewable generation but increase the building energy demand.

Furthermore, with respect to the transition from renewable surplus toward demand shortage region, the cumulative distribution probability for the stochastic uncertainty scenario is higher than the deterministic scenario, i.e., 89.4% (stochastic) vs. 87.0% (deterministic) for the office building, 93.5% (stochastic) vs. 85.8% (deterministic) for the hotel building, 75.9% (stochastic) vs. 47.7% (deterministic) for the residential building, and 71.9% (stochastic) vs. 68.9% (deterministic) for the commercial building. This indicates that, compared to stochastic uncertainty analysis, the deterministic analysis will overestimate the building energy performance with a lower cumulative distribution probability for the transition from renewable surplus to demand shortage region. Therefore, stochastic uncertainty analysis is quite necessary to avoid performance overestimation.

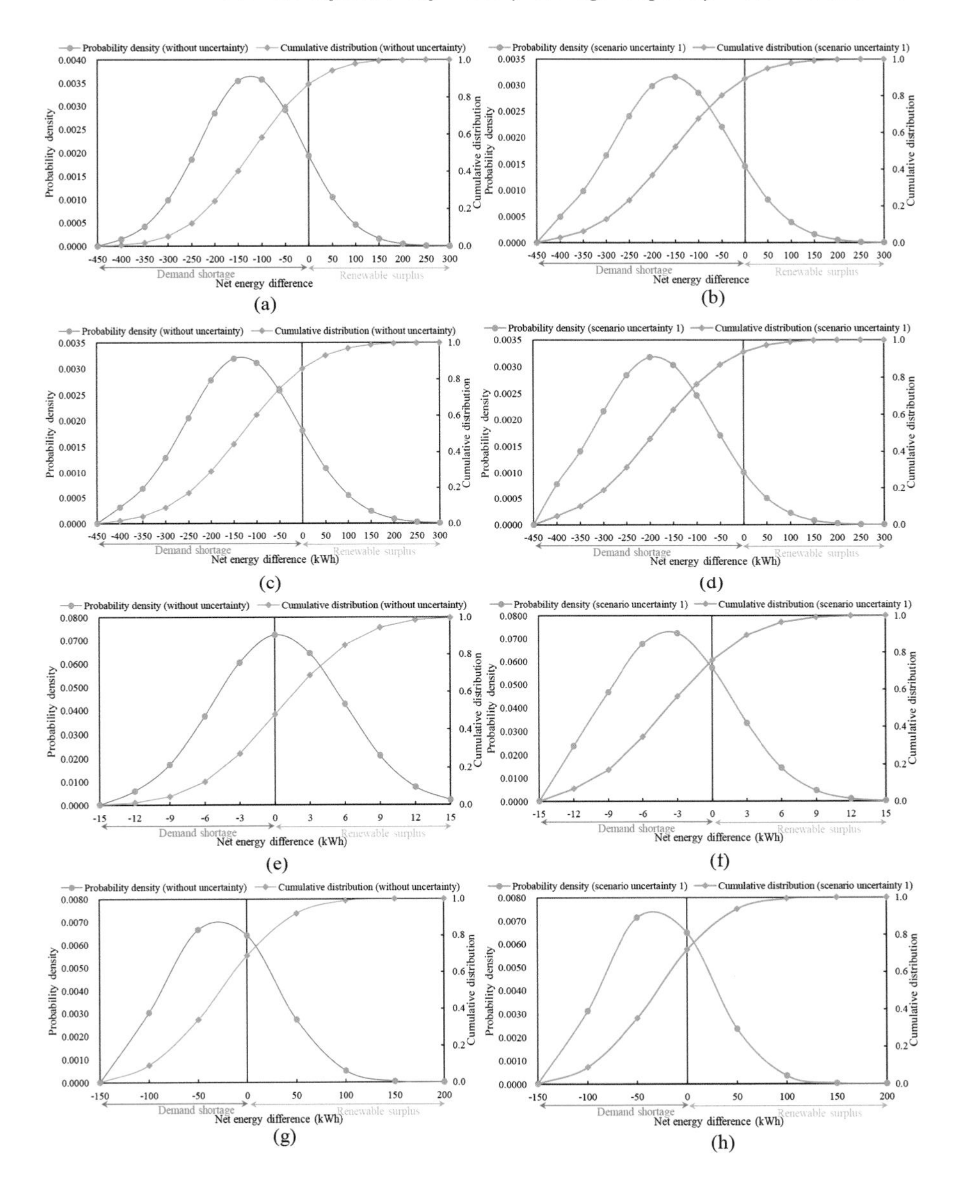

Figure 2.7 Probability density and cumulative distribution of net energy difference between renewable and demand for the deterministic scenario: (a) office, (c) hotel, (e) residential, and (g) commercial building; and scenario with parameter uncertainty: (b) office, (d) hotel, (f) residential, and (h) commercial building.

(Note: The negative value in net energy difference indicates that the demand is higher than the renewable energy. The positive value in net energy difference indicates that renewable energy is higher than the demand.)

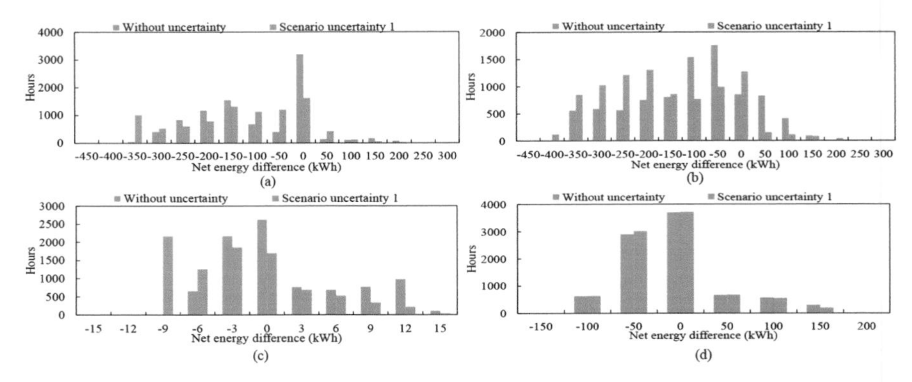

Figure 2.8 Distribution for time duration of net energy difference for: (a) high-rise office; (b) high-rise hotel; (c) residential building; (d) commercial building.

Figure 2.8 shows the frequency of hours for net energy difference. As shown in Figure 2.8, the scenario uncertainty shows longer hours in the negative net energy difference compared to the deterministic scenario. For instance, compared to the deterministic case with the total hours for the negative net energy difference at 5085, 6571, 2843, and 3524 hours for the high-rise office, the high-rise hotel, the residential, and the commercial building, the total hours for the negative net energy difference for the stochastic scenarios are 6554, 7106, 5275, and 3636 hours for the high-rise office, the high-rise hotel, the residential, and the commercial building, respectively. This indicates that considering scenario uncertainties will increase the demand shortage periods.

2.4.2 Identification of battery storage capacity for deterministic and stochastic scenarios with uncertainty

Based on the cumulative distribution probability of net energy difference between renewable generations and demands, the uncertainty-based risk acceptance methodology was adopted to size the battery storage capacity in deterministic and stochastic uncertainty scenarios. In this study, the first priority for the battery was to cover building energy demand, and the function of storing surplus renewable energy was given with sub-priority. Therefore, the battery storage capacity is sized based on the negative net energy difference.

Figure 2.9 demonstrates the principle for battery storage capacity sizing in different cumulative distribution probabilities. As shown in Figure 2.9, with the increase of the accepted risk, the absolute value of demand shortage decreases gradually in both deterministic and stochastic scenarios. As listed

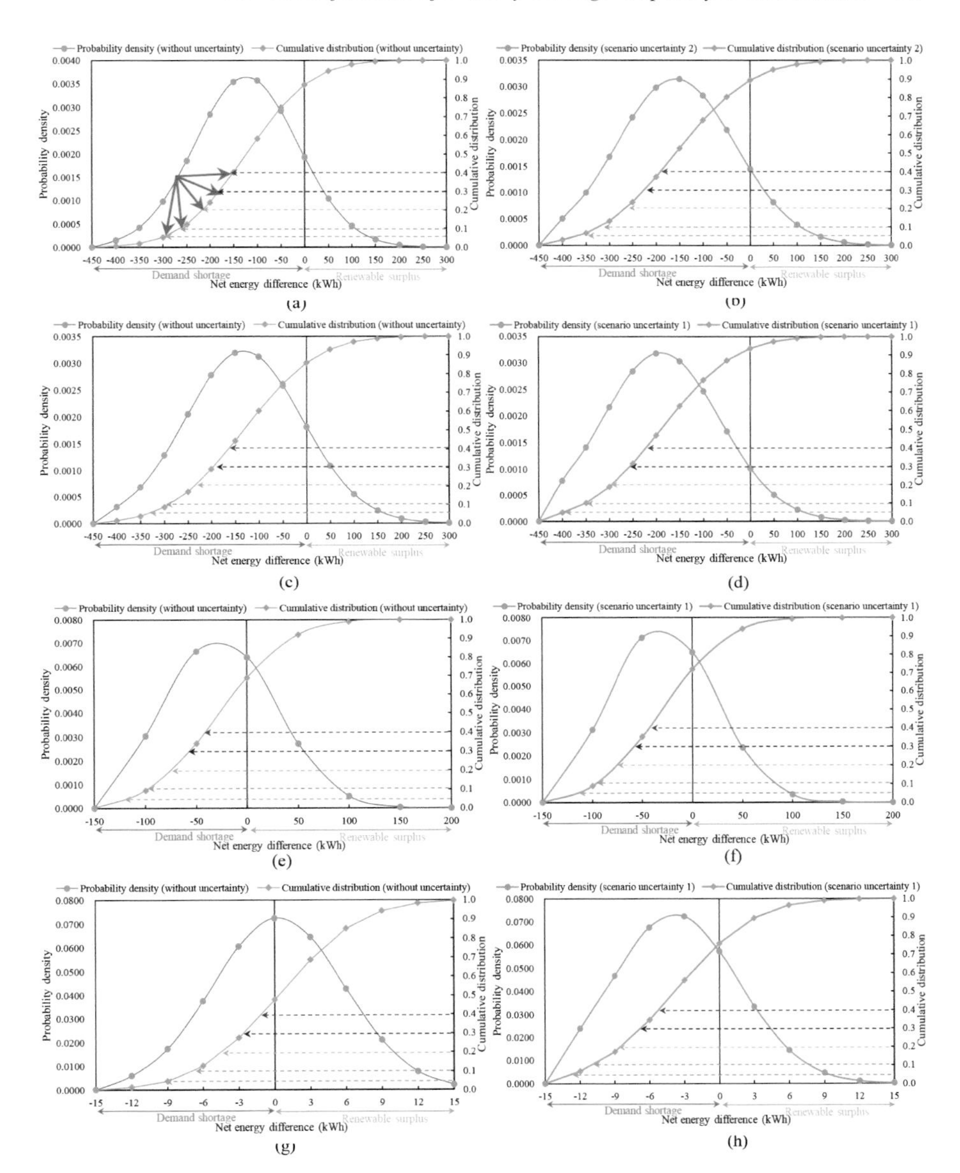

Figure 2.9 Determination for battery storage capacity: deterministic scenario: (a) office, (c) hotel, (e) commercial building, and (g) residential building; stochastic case with scenario uncertainty: (b) office, (d) hotel, (f) commercial building, and (h) residential building.

Table 2.5 Battery storage capacity for deterministic scenario.

Accepted risk	**5%**	**10%**	**20%**	**30%**	**40%**
Office battery (kWh)	300	260	220	180	150
Hotel battery (kWh)	330	290	240	190	160
Commercial battery (kWh)	120	100	80	60	40
Residential battery (kWh)	9	7	4.5	2	1

Table 2.6 Battery storage capacity for stochastic scenario with uncertainty.

Accepted risk	**5%**	**10%**	**20%**	**30%**	**40%**
Office battery (kWh)	350	320	260	230	190
Hotel battery (kWh)	400	350	300	250	220
Commercial battery (kWh)	110	95	75	55	45
Residential battery (kWh)	13	11	9	7	6

in Table 2.5, the increase of the accepted risk from 5% to 40% will lead to a decrease in the battery capacity from 300 to 150 kWh (office), from 330 to 160 kWh (hotel), from 120 to 40 kWh (commercial building), and from 9 to 1 kWh (residential building). The underlying mechanism is that, with the increase of the accepted risk, the battery system cannot cover more building energy demands. Meanwhile, except for the commercial building, compared to the deterministic scenario, the stochastic scenario with uncertainty, as listed in Table 2.6, shows a much higher battery storage capacity. For instance, when the accepted risk is 5%, the stochastic scenario with uncertainty will increase the battery storage capacity from 300 to 350 kWh (office), from 330 to 400 (hotel), and from 9 to 13 kWh (residential building). The underlying mechanism is that, compared to the deterministic scenario, the stochastic scenario with uncertainty shows a much higher time duration for the negative net energy difference, as shown in Figure 2.8 (Section 2.3.3.1). It is noteworthy that the battery storage capacity of the commercial building is slightly lower for the stochastic scenario with uncertainty than the deterministic scenario unless the accepted risk approaches 40%. The reason is due to the slight difference in the time-duration distribution of net energy difference between deterministic and stochastic scenarios, as shown in Figure 2.9(e) and (f), and the higher average absolute value of net energy difference for stochastic scenario (−30.1 kWh) than the deterministic scenario (−27.3 kWh).

2.5 Conclusion

With respect to the district energy community with stochastic uncertainty on multi-diversified energy demands and distributed renewable generations, a generic approach for battery storage sizing is critical in terms of demand coverage, grid stability, and reliable power supply with the integration of intermittent renewable generations. In this study, a district energy community was formulated, consisting of distributed renewable systems (i.e., rooftop PVs and BIPVs), different types of buildings (a 30-floor office, a 30-floor hotel, a 2-floor residential building, and a 2-floor commercial building), and static battery storages in each building. Using the component-wise Metropolis–Hastings algorithm, a two-dimensional Markov chain Monte Carlo (MCMC) multivariate distribution was adopted to quantify scenario uncertainty with the bivariate distribution. Through feature extraction and classification, data-driven models were structured and then trained by a cross-entropy algorithm for accurate predictions on building energy demands and renewable generations under stochastic uncertainties. In terms of the accuracy and reliability of supervised learning trained data-driven models, statistical indicators were adopted, such as coefficient of determination (R^2), the mean absolute error (MAE), the root mean square error (RMSE), the normalized mean bias error (NMBE), and the coefficient of variation or the root mean square error (CV-RMSE). Based on the dynamic difference between distributed renewable generations and building energy demands, a risk-based battery sizing approach was adopted to identify the storage capacity for each battery. The comparative analysis between deterministic and stochastic scenarios demonstrates the effectiveness of the proposed techniques. Remarkable conclusions are drawn as follows:

- The supervised learning-based surrogate models are reliable and accurate for predictions of energy demands in district buildings and renewable energy generations. The feature extraction and classification on energy demands in different building types shows that the RMSEs are 14.49, 5.45, and 0.58 kW for the hotel, the office, and the residential building. Meanwhile, the NMBEs are –0.77%, –0.29%, and 0.97%, and the CV-RMSEs are 7.89%, 3.13%, and 21.07% for the hotel, the office, and the residential building, respectively. Furthermore, in terms of the prediction of renewable generations, the RMSEs are 9.48, 1.26, and 0.43 kW for office, commercial, and residential buildings. Meanwhile, the NMBEs are 0.067%, –0.153%, and 1.35%, and the CV-RMSEs are 18.6%, 3.21%, and 14.2% for the office, commercial, and residential buildings, respectively. The developed surrogate models can meet the requirement defined in ASHRAE Guideline 14, indicating the applicability and accuracy of dynamic performance prediction.
- By driving the well-trained surrogate models, uncertainty-based predictions on building energy demands and distributed renewable generations indicate that, compared to the deterministic scenario, the

stochastic uncertainty-based scenario shows higher building energy demands and lower distributed renewable generations. From the histogram on electric demand, compared to the deterministic scenario, the stochastic scenario with uncertainty will increase the time duration for electrical demand at a higher value but decrease the time duration for renewable generation at a higher value.

- The net energy difference between distributed renewable generations and district demands indicates that, compared to the stochastic uncertainty analysis, the deterministic analysis will overestimate the building energy performance with a lower cumulative distribution probability for the transition from renewable surplus to demand shortage regions. Furthermore, the mean value of the net energy difference for the deterministic solution is much lower than the stochastic solution.
- By adopting the uncertainty-based risk acceptance methodology for battery storage sizing, the battery storage capacity indicates that, compared to the deterministic scenario, the stochastic scenario with uncertainty shows a much higher battery storage capacity. When the accepted risk is 5%, the stochastic scenario with uncertainty will increase the battery storage capacity from 300 to 350 kWh (office), from 330 to 400 kWh (hotel), and from 9 to 13 kWh (residential building). Furthermore, the increase of the accepted risk from 5% to 40% will lead to a decrease in the battery capacity from 300 to 150 kWh (office), from 330 to 160 kWh (hotel), from 120 to 40 kWh (commercial building), and from 9 to 1 kWh (residential building).

With the comprehensive consideration of stochastic uncertainties in distributed renewable generations and multi-diversified energy demands, supervised learning trained surrogate models were developed to conduct the dynamic performance predictions. The uncertainty-based risk acceptance methodology adopted for battery storage capacity sizing was based on the cumulative distribution probability of net energy difference between renewable generation and energy demand. The proposed approach can be applied for energy planning on battery storage systems with scenario uncertainties, providing preliminary design and operation strategies for energy management in a large-scale district energy community. Compared to the deterministic case study, the uncertainty-based approach can improve the reliability and robustness of the district energy community to improve renewable penetration and reduce grid reliance.

References

[1] P Nejat, F Jomehzadeh, MM Taheri, M Gohari, MZA Majid. A global review of energy consumption, CO2 emissions, and policy in the residential sector (with an overview of the top ten CO2 emitting countries). Renewable and Sustainable Energy Reviews 2015, 43, 843–862.

[2] EnergyPlus. https://energyplus.net/

[3] TRNSYS. http://www.trnsys.com/

[4] DesignBuilder. https://designbuilder.co.uk/

[5] Y Zhou, S Zheng. Machine-learning based hybrid demand-side controller for high-rise office buildings with high energy flexibilities Applied Energy 2020, 262, 114416.

[6] S Fathi, R Srinivasan, A Fenner, S Fathi. Machine learning applications in urban building energy performance forecasting: A systematic review. Renewable and Sustainable Energy Reviews 2020, 133, 110287.

[7] K Amasyali, NM El-Gohary. A review of data-driven building energy consumption prediction studies. Renewable and Sustainable Energy Reviews 2018, 81, 1192–1205.

[8] Y Zhou. Artificial neural network based smart aerogel glazings in low-energy buildings — A state-of-the-art review. iScience 2021, Volume 24, Issue 12, pp 1–26. https://doi.org/10.1016/j.isci.2021.103420

[9] C Zhang, J Li, Y Zhao, T Li, Q Chen, X Zhang. A hybrid deep learning-based method for short-term building energy load prediction combined with an interpretation process. Energy and Buildings 2020, 225, 110301.

[10] SØ Jensen, A Marszal-Pomianowska, R Lollini, W Pasut, A Knotzer, P Engelmann, A Stafford, G Reynders. IEA EBC annex 67 energy flexible buildings. Energy and Buildings 2017, 155, 15 25–34.

[11] Y Zhou, S Zheng, Z Liu, T Wen, Z Ding, J Yan, G Zhang. Passive and active phase change materials integrated building energy systems with advanced machine-learning based climate-adaptive designs, intelligent operations, uncertainty-based analysis and optimisations: A state-of-the-art review. Renewable and Sustainable Energy Reviews 2020, 130, 109889.

[12] Y Zhou, S Cao. Energy flexibility investigation of advanced grid-responsive energy control strategies with the static battery and electric vehicles: A case study of a high-rise office building in Hong Kong. Energy Conversion and Management 2019, 199, 111888.

[13] JP Deane, BPÓ Gallachóir, EJ McKeogh. Techno-economic review of existing and new pumped hydro energy storage plant. Renewable and Sustainable Energy Reviews 2010, 14, 4, 1293-1302.

[14] J Liu, H Yang, Y Zhou. Peer-to-peer trading optimizations on net-zero energy communities with energy storage of hydrogen and battery vehicles. Applied Energy 2021, 302, 117578.
[15] Y He, Y Zhou, Z Wang, J Liu, Z Liu, G Zhang. Quantification on fuel cell degradation and techno-economic analysis of a hydrogen-based grid-interactive residential energy sharing network with fuel-cell-powered vehicles. Applied Energy 2021, 303, 117444.
[16] Y Zhou, S Cao, JLM Hensen, PD Lund. Energy integration and interaction between buildings and vehicles: A state-of-the-art review. Renewable and Sustainable Energy Reviews 2019, 114, 109337.
[17] Y Yang, S Bremner, C Menictas, M Kay. Battery energy storage system size determination in renewable energy systems: A review. Renewable and Sustainable Energy Reviews 2018, 91, 109–125.
[18] Y Zhou, S Zheng, G Zhang. Machine-learning based study on the on-site renewable electrical performance of an optimal hybrid PCMs integrated renewable system with high-level parameters' uncertainties. Renewable Energy 2020, 151, 403-418.
[19] W Tian, Y Heo, P De Wilde, Z Li, D Yan, CS Park, A review of uncertainty analysis in building energy assessment. Renewable and Sustainable Energy Reviews 2018, 93, 285–301.
[20] D Coppitters, W De Paepe, F Contino. Robust design optimization of a photovoltaic-battery-heat pump system with thermal storage under aleatory and epistemic uncertainty. Energy 2021, 229, 120692.
[21] Y Zhou, S Zheng, G Zhang. A state-of-the-art-review on phase change materials integrated cooling systems for deterministic parametrical analysis, stochastic uncertainty-based design, single and multi-objective optimisations with machine learning applications. Energy and Buildings 2020, 220, 110013.
[22] ASHRAE. ASHRAE Guideline 14: Measurement of Energy, Demand and Water Savings American Society of Heating, Refrigeration and Air Conditioning Engineers, Atlanta, GA, USA, 2014.
[23] JH Mcclellan, RW Schafer, MA Yoder. Signal Processing First: International Edition, Pearson Schweiz Ag, Zug, Switzerland, 2003.

Chapter 3

Design and Development of Solar-Powered Hybrid Energy Bank

Raghu Chandra Garimella*[1], Siva Rama Krishna Madeti[2], T. Bhavani Shankar[3], K. Raghavendra Nayak[3], M. Kumar[3], Gaurav Saini[4], and Krishna Kumar[5]

*[1]Associate Professor, Department of Electrical & Electronics Engineering, Methodist College of Engineering and Technology, India
[2]Assistant Professor, University of Santiago de Chile, Chile
[3]Methodist College of Engineering and Technology, India,
[4]Assistant Professor, Department of Mechanical Engineering, Harcourt Butler Technical University Kanpur, India
[5]Research Scholar, IIT Roorkee, India
Corresponding authors: raghuchandhra@gmail.com, ramakrishna.iitroorkee@gmail.com

Abstract

Due to their features of extending the operating duration for portable electronic gadgets and mobile phones, which are part of modern life, power banks are growing significantly. In a mobile phone, the onshore battery can provide power for a few hours and then gets died out over time. Therefore, a power bank is needed to extend the operation of electronic gadgets like mobile phones. However, charging electronic gadgets or mobile phones using non-conventional energy resources, such as solar panels, will be helpful in remote areas and beneficial to the environment. This chapter describes the design and development of the solar-powered hybrid energy bank for a mobile phone/electronic gadget (Granted Patent Ref.: AU2020100836A4; AU2020100836; IN327433).

3.1 Introduction

In the present research, a conventional power bank is redesigned as a hybrid system using polycrystalline material-based solar panels to facilitate the most efficient and renewable-energy-based unlimited powered energy bank, which fulfills the need of an hour in the society (Garimella *et al.*, 2020a, 2020b). The solar-powered hybrid energy bank system comprises solar panels composed of solar cells, absorbing solar radiation and converting it into electrical energy. In addition, the solar-powered hybrid energy bank system also comprises conventional charging mechanism port. Further, the system uses a charge control module to control the charging/discharging mechanism of the Li-ion battery (2600 mAh), which is mounted in a casing. Further, the electrical energy generated from solar panels or external sources of electricity through the conventional charging mechanism port is transferred to the charge controlling module, where the built-in Li-ion battery gets charged. Subsequently, the output of the Li-ion battery is connected to the chopper circuit or DC/DC boost module, which is a part of the solar-powered hybrid energy bank system. The chopper circuit, a DC equivalent of the AC transformer, boosts the lower-level voltage to the required load range. The proposed connection diagram of the solar-powered hybrid energy bank devised is depicted in Figure 3.1.

Currently, the charge control module is used to provide circuit protection, which regulates the overcharge and discharge phenomena of the Li-ion battery. The complementary metal oxide semiconductor (CMOS) based pulse frequency modulation (PFM) controlled step-up switching DC/DC converter module (CE8301), generally called a boost chopper module, is used to step up the input lower voltages to the required level of voltage, about 5 V in this invention.

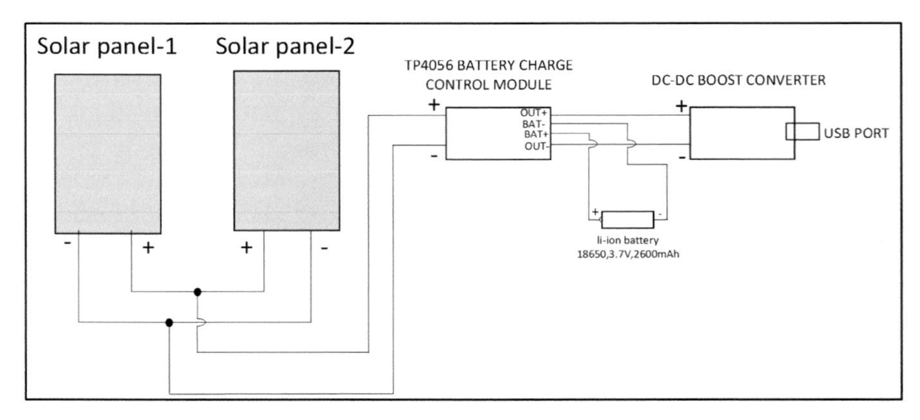

Figure 3.1 Proposed connection diagram of the solar-powered hybrid energy bank devised.

3.2 Literature Review

A power bank is a portable device used to charge electronic gadgets and is shown in Figure 3.2 (Joshi, 2015; Leschin *et al.*, 2007; Smith *et al.*, 2013). Power banks may generally be selected based on mAh rating (Adetona *et al.*, 2020; Chao and Tai, 2017; Divya *et al.*, 2018; Pitao Jr *et al.*, 2019; Suen *et al.*, 2017). Typically, the conventional power banks are designed with a combination of a charge control module, rechargeable battery or batteries, and a DC/DC boost module (Chung and Trescases, 2017; Droppo *et al.*, 2005; Jacobs, 2007; Tamai and Aldrich III, 2001; Wu, 2010; Yilmaz and Krein, 2012). Many technical problems were identified in the conventional power bank systems (Auer et al., 2012; Gunawan *et al.*, 2020; Liu, 2010). A conventional power bank is not everlasting, and they are needed to be charged to use continuously (Omer, 2008; Ter-Gazarian, 1994). The major drawback of conventional power banks is the characteristic feature of utilizing electrical energy to charge the battery and/or batteries of them (Bailey, 2019; Kindeskog and Pettersson, 2016; Rhodes, 2016).

Therefore, there is a need for an improved system despite the existence of conventional power banks, which cannot be charged in electrically isolated areas (Chauhan and Saini, 2016; de Souza Ribeiro *et al.*, 2010; Fernandez *et al.*, 2005; Lal *et al.*, 2011; Ribeiro *et al.*, 2012). The present research relates to the design and development of a solar-powered hybrid energy bank system, which may be used to charge battery of a rechargeable electronic gadget through either solar panels and/or a conventional charging mechanism. Some

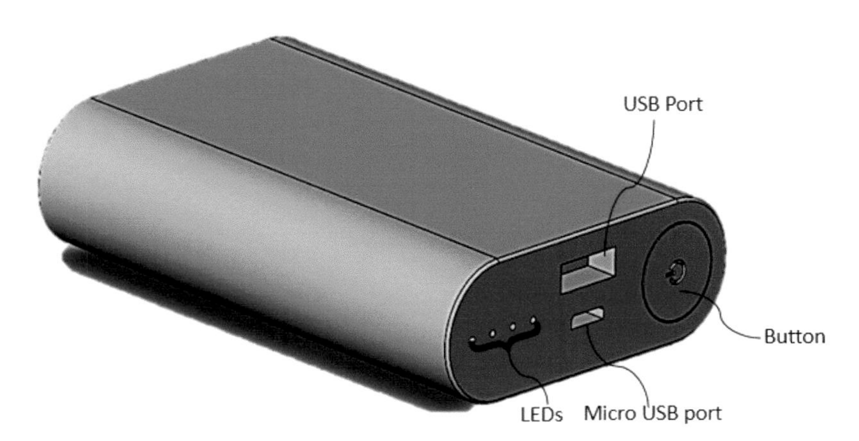

Figure 3.2 Conventional power bank unit.

of the advantages of the present system include, but are not limited to, the following:

- Use of renewable energy in addition to the conventional charging mechanism, i.e., a built-in hybrid mechanism that allows to charge from both the solar radiation and the external source of electricity.
- Employment of polycrystalline material-based solar panels that are cost-effective, thinner, very efficient, and have a reflectance efficiency is just about 1%–2%.
- The best-suited system for electrically isolated areas, which do not have electricity access, has very high thermal conductivity.

3.3 Material and Methods

3.3.1 Solar panels

The solar panels are one of the charging sources for the in-built Li-ion battery incorporated in the embodiment of the energy bank. A casing is provided to the Li-ion battery for being properly mounted in the energy bank. The solar panels are constructed by combining multiple layers of silicon-based solar cells made of polycrystalline material that works on the basic principle of absorbing radiation from the sun and converting the solar radiation into useful electricity. Silicon-based solar cells are covered with glass lamination. The solar cells provide glass lamination to avoid any possible damage. Figure 3.3 depicts the isometric view of solar panels devised.

The complete specifications and ratings of the solar panels are elucidated in Table 3.1.

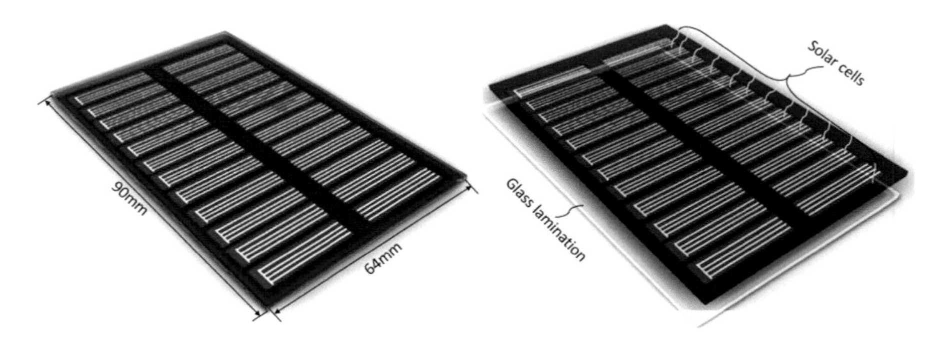

Figure 3.3 Polycrystalline solar panel.

Table 3.1 Specifications of solar panels devised.

Parameter	Range
Output voltage (max.)	5 V
Output current (max.)	100 mA
Rated power	0.5 W
Material	Polycrystalline
Dimensions	90 × 64 × 2 mm

3.3.2 Control circuits

The solar power bank consists of a charge control module and a DC–DC step-up module. The basic function of a charge control module is to avoid the overcharge or complete discharge phenomena in the Li-ion battery, which prevents the battery's improper functioning. The charge control module circuit mainly comprises Li-ion battery charger IC (TP4056), Li-ion battery protection IC (DW01A), Dual N-Channel MOSFET (FS8205A), Micro-B USB pinout that supports a conventional charging mechanism, and two LED lights. The Li-ion battery charger IC (TP4056) is provided to attain complete linear constant current with the help of the internal PMOSFET architecture, which prevents the reverse current flow. The integral thermal feedback circuit in Li-ion battery charger IC (TP4056) regulates the charge current for limiting the temperature rise during high power operation. Figure 3.4 illustrates

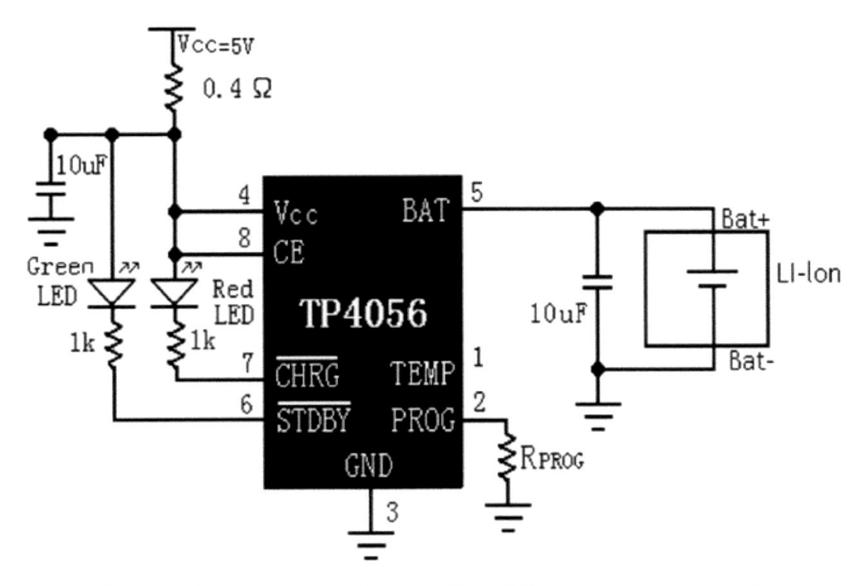

Figure 3.4 External architecture of TP4056 IC (Anonymous, 2018).

Table 3.2 Pin description of TP4056 IC.

Pin number	Description
Pin 1	TEMP
Pin 2	PROG
Pin 3	GND
Pin 4	Vcc
Pin 5	BAT
Pin 6	STDBY
Pin 7	CHRG
Pin 8	CE

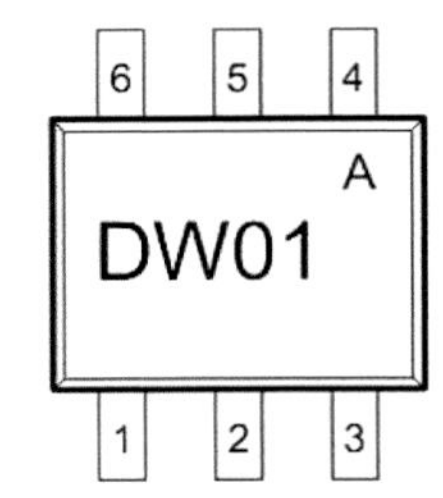

Figure 3.5 External architecture of DW01A IC (Anonymous, 2020a, p. 01, 2020b).

Table 3.3 Pin description of DW01A IC.

Pin number	Description
Pin 1	MOSFET gate connection for discharge control (OD)
Pin 2	Input pin for current sense charger detect (CS)
Pin 3	MOSFET gate connection for charge control (OC)
Pin 4	Test pin to reduce delay time (TD)
Pin 5	Power supply (Vcc)
Pin 6	Ground (GND)

the external architecture (pin diagram) of Li-ion battery charger IC (TP4056), and Table 3.2 provides the pin description of TP4056 IC.

Further, the charge control module circuit comprises a Li-ion battery protection IC (DW01A), which has six (06) pins to prevent the Li-ion battery from being overcharged and reduce the life cycle. Figure 3.5 illustrates the external architecture (pin diagram) of Li-ion battery protection IC (DW01A), and Table 3.3 provides the pin description of DW01A IC.

The Dual N-Channel MOSFET (FS8205A) is an 8-pin IC used to protect the Li-ion battery. FS8205 switches off the Li-ion battery during the occurrence of complete discharge. Subsequently, Micro-B USB pinout is employed to provide the conventional charging mechanism, the traditional way of charging, in addition to the solar-panels-based charging mechanism. This particularly makes the unit a hybrid system.

3.3.3 Battery

An electrical battery involves in electrochemical energy conversion that delivers charge to the load based on a load demand (Badwal *et al.*, 2014; Dunn *et al.*, 2011; Hou *et al.*, 2011; Pires *et al.*, 2014). In the solar-powered hybrid energy bank system, a Li-ion battery made up of secondary cells (rechargeable) type is employed. Further, the Li-ion battery with a positive terminal and a negative terminal is equipped with a casing for mounting

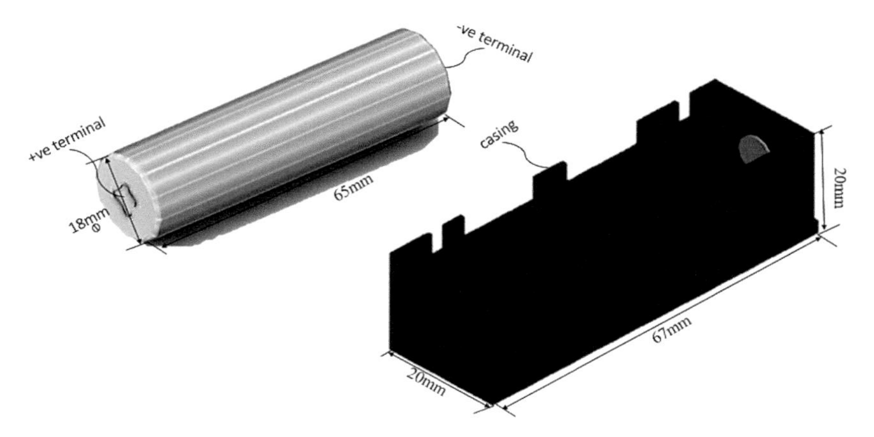

Figure 3.6 Isometric view of a Li-ion battery and the casing.

Table 3.4 Specifications of Li-ion battery employed.

Parameter	**Range**
Voltage	3.2–4.2 V
mAh rating	2600 mAh
Max discharge current	4400 mA
Min discharge current (standard discharge)	440 mA
Max charge current	2000 mA
Dimensions	Diameter: 18 mm; Height: 65 mm
Weight	41 g
Type of battery	Lithium-ion (rechargeable)

properly in the energy bank. Figure 3.6 portrays the isometric view of a Li-ion battery, and it is casing.

The Li-ion battery is charged through solar panels, and the charging process will be controlled with the help of a charge control module circuit. Moreover, the charging of Li-ion battery may also happen with Micro-B USB pinout through a conventional charging mechanism using external sources of electricity. The Li-ion battery employed in the solar-powered hybrid energy bank system has about 2600-mAh capacity with two (02) 100-mA rated solar panels connected in parallel. The complete specifications and ratings of the Li-ion battery are elucidated in Table 3.4.

3.3.4 DC/DC boost module

The boost module provides a stable 5-V supply to the connected loads with a maximum of 1-A current at the output side. The input of the boost-up module is connected to the solar panel and/or the charge control module. The major components of the DC/DC boost module are 47-μH rated on-board inductor, CMOS-based PFM control step-up switching DC/DC converter module IC (CE8301), Type-A USB Pin, and an LED. The on-board inductor is operated along with the CMOS-based PFM control step-up switching DC/DC converter module IC (CE8301) to obtain a required power level based on load demand as of Li-ion capacity. The CE8301 of the DC/DC boost module comprises an oscillator and a comparator. The PFM control automatically switches the duty cycle based on load. The DC/DC converter module IC (CE8301) has an on-board metal oxide semiconductor field effect transistor (MOSFET), which turns off the system while being overcharged to prevent the Li-ion battery from being damaged. Type-A USB Pin is provided in the DC/DC boost module circuit for charging an electronic gadget through an external source of electricity. An LED indicates the connection of an external device (an electronic gadget) to the DC/DC boost module circuit. Figures 3.7 and 3.8 illustrate an external architecture

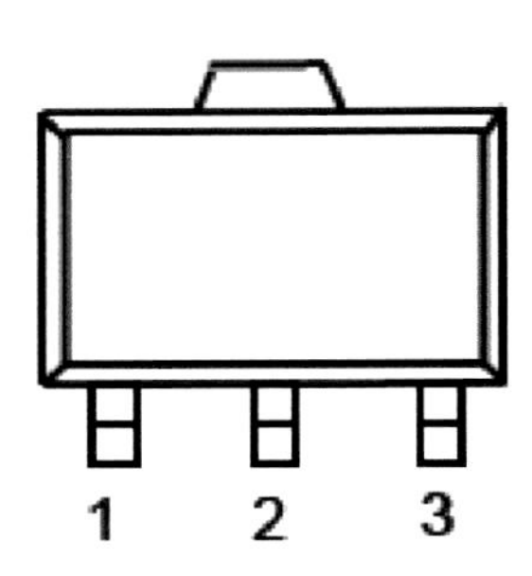

Figure 3.7 External architecture (pin diagram) of CMOS-based PFM control step-up switching DC/DC converter module IC (CE8301) (Anonymous, 2020c).

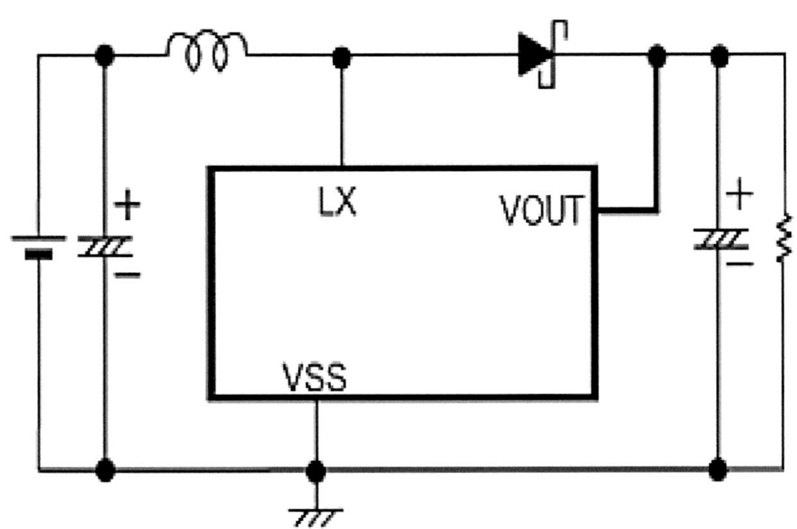

Figure 3.8 Internal architecture of IC CE8301 (Anonymous, 2020c).

Table 3.5 Specifications and ratings of the DC/DC boost module.

Components	Range	Quantity
CE8301 IC	–	01
Capacitor	0.1 μF	01
Capacitor	100 μF	01
Inductor	47 μH	01
LED	–	01

(pin diagram) and an internal architecture of CMOS-based PFM control step-up switching DC/DC converter module IC (CE8301).

The complete specifications and ratings of the DC/DC boost module are elucidated in Table 3.5.

3.4 Investigations

The solar-powered hybrid energy bank system may be associated with an electronic gadget or any other device performing various functions. The functions may include, but are not limited to, charging a battery of the gadget/device through solar energy; charging a battery of the gadget/device through an external source of electricity; charging a battery of the gadget/device through both the combination of solar energy and external source of electricity; and providing DC power to external loads. Figure 3.9 depicts the

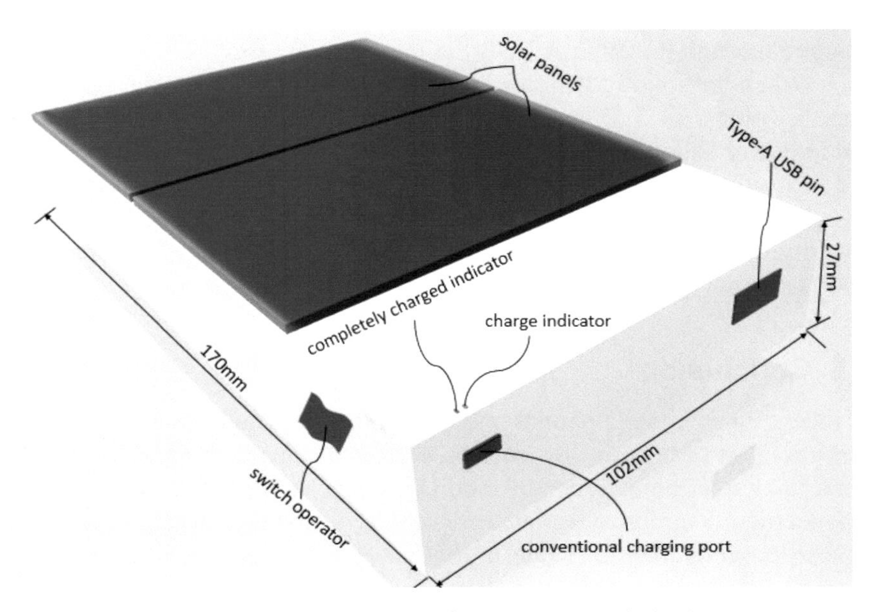

Figure 3.9 Isometric view of the energy bank devised.

isometric view of the energy bank devised that illustrates a partial perspective view.

The externally disposed around the case of the energy bank is designed using a lesser weight and high-quality plastic material made up of poly vinyl chloride (PVC). This helps to provide good insulation to the Li-ion battery and the entire circuitry from the electricity and the temperatures as well. Therefore, the present research offers a hybrid energy bank that is efficient, economical, effective, and flexible for the provision of portable electricity to various applications using solar and conventional mechanisms.

3.5 Results and Discussion

The solar radiation tapped from the sun is collected with the help of laminated solar panels, which are fabricated by integrating several silicon-based solar cells made up of polycrystalline material. The tapped radiation through the solar panels is converted into electrical energy using the photovoltaic effect. Further, the charge control circuit, which is connected in series to the solar panels and Li-ion battery, regulates the power ratings and helps the Li-ion battery to charge through the renewable energy and/or conventional charging mechanism through Micro-B USB pinout. Subsequently, the DC/DC boost converter, which is connected in series with the charge control circuit and Li-ion battery, intensifies the voltage ratings as per load requirements, about 5 V in this invention. For instant electrical load, an electronic gadget may be charged through the Type-A USB Pin placed in the DC/DC boost converter

When the solar radiations fall on the solar panels, the solar panels generate a voltage of 5 V and energize the charge control module. Further, a red indicator shows the flow of electricity through the embedded panels, and the charging module starts charging the battery. When the battery is fully charged, it automatically disconnects the battery from the charging module, ensuring it is not overcharged. Figure 3.10 depicts the designed and developed solar-powered hybrid energy bank system model.

3.6 Conclusion

In recent times, mobile phones and many electronic gadgets have become an essential part of human life. Further, conventional power banks were developed to charge the batteries of mobile phones and/or electronic gadgets. In the electrically remote or isolated areas, the conventional power banks are not able to provide electricity due to the insufficiency of charging themselves.

Figure 3.10 Designed and developed model of a solar-powered hybrid energy bank system.

Therefore, the solar-powered hybrid energy bank system is developed as the next alternative to charge the mobile phones and/or electronic gadgets. The solar-powered hybrid energy bank system charges itself from the renewable solar radiations and under artificial lights as well. Draining out of the solar power bank is, therefore, almost impossible.

3.7 Acknowledgment

The authors would like to acknowledge the management of Methodist College of Engineering and Technology (MCET), Osmania University, for permitting the work. Opinions expressed in the paper are of the author and need not be of the institution.

3.8 Funding Sources

This research received complete funding from Methodist College of Engineering and Technology (MCET), Osmania University, to design and develop the solar-powered hybrid energy bank. The author is so thankful to the Management of MCET for their extended support and encouragement during the research.

References

[1] Adetona, S., Ige, M., Salawu, R., 2020. A mechanically powered 3500 mAh mobile phones power-bank. J. Eng. Res. 25, 88–98.

[2] Anonymous, 2020a. DW01A datasheet, Pinout application circuits one cell Lithium-ion/polymer battery protection IC [WWW Document]. URL http://www.datasheetdir.com/DW01A+Battery-Protectors-Monitors (accessed 11.12.20).

[3] Anonymous, 2020b. General description ordering information features applications [WWW Document]. URL http://webcache.googleusercontent.com/search?q=cache:OlFWROHYVfIJ:hmsemi.com/downfile/DW01A.PDF+&cd=1&hl=en&ct=clnk&gl=in (accessed 11.12.20).

[4] Anonymous, 2020c. CE8301 Series. Introduction. Features. Ordering Information. Applications small package PFM control Step-Up Dc/Dc converter - PDF free download [WWW Document]. URL https://docplayer.net/14478682-Ce8301-series-introduction-features-ordering-information-applications-small-package-pfm-control-step-up-dc-dc-converter.html (accessed 11.12.20).

[5] Anonymous, 2018. TP4056 Lithium-ion battery charger - circuit, 18650 battery charging. Electron. Hub. URL https://www.electronicshub.org/tp4056-lithium-ion-battery-charger/ (accessed 11.12.20).

[6] Auer, J., Keil, J., Stobbe, A., AG, D.B., Mayer, T., 2012. State-of-the-art electricity storage systems. Dtsch. Bank DB Res.

[7] Badwal, S.P., Giddey, S.S., Munnings, C., Bhatt, A.I., Hollenkamp, A.F., 2014. Emerging electrochemical energy conversion and storage technologies. Front. Chem. 2, 79 pp 1–28.

[8] Bailey, G.B., 2019. Wind or water driven electric generator used to recharge Li-ion batteries in field applications.

[9] Chao, H.-C., Tai, L.D., 2017. Portable powerbank with integrated promotional-video capabilities and methods of use.

[10] Chauhan, A., Saini, R.P., 2016. Techno-economic feasibility study on integrated renewable energy system for an isolated community of India. Renew. Sustain. Energy Rev. 59, 388–405.

[11] Chung, S., Trescases, O., 2017. Hybrid energy storage system with active power-mix control in a dual-chemistry battery pack for light electric vehicles. IEEE Trans. Transp. Electrification 3, 600–617.

[12] de Souza Ribeiro, L.A., Saavedra, O.R., De Lima, S.L., De Matos, J.G., 2010. Isolated micro-grids with renewable hybrid generation: The case of Lençóis island. IEEE Trans. Sustain. Energy 2, 1–11.

[13] Divya, M., Saravanan, K., Balaji, G.N., Pandian, S.C., 2018. Light weight & low cost power bank based on LM7805 regulator for hand held applications. International Journal of Latest Technology in Engineering Management & Applied Science (IJLTEMAS) 7, 201–205.

[14] Droppo, G.W., Schienbein, L.A., Harris, B.E., Hammerstrom, D.J., 2005. DC to DC converter and power management system.

[15] Dunn, B., Kamath, H., Tarascon, J.-M., 2011. Electrical energy storage for the grid: a battery of choices. Science 334, 928–935.

[16] Fernandez, H., Martinez, A., Guzman, V., Gimenez, M., 2005. An experimental and training platform for uninterrupted power in isolated locations: Wind turbine-electric generator-battery bank (TP-WT-UPS), in: 2005 European Conference on Power Electronics and Applications. IEEE, p. 8.

[17] Garimella, R.C., Kethavath, R.N., Madeti, S.R., Methri, K., T, B.S., 2020a. Portable hybrid power bank. AU2020100836A4.

[18] Garimella, R.C., Kethavath, R.N., T, B.S., Methri, K., 2020b. Intellectual property India [WWW Document]. URL http://ipindiaservices.gov.in/PatentSearch/PatentSearch//ViewApplicationStatus (accessed 11.12.20).

[19] Gunawan, V., Christiardy, N., Irawan, A.P., 2020. Concept design of high interface powerbank, in: IOP Conference Series: Materials Science and Engineering. IOP Publishing, p. 012089.

[20] Hou, J., Shao, Y., Ellis, M.W., Moore, R.B., Yi, B., 2011. Graphene-based electrochemical energy conversion and storage: Fuel cells, supercapacitors and lithium ion batteries. Phys. Chem. Chem. Phys. 13, 15384–15402.

[21] Jacobs, J.K., 2007. Battery controller and method for controlling a battery.

[22] Joshi, A., 2015. Low cost, portable and extendable power bank. Int. J. Sci. Eng. Res. 6, 87–89.

[23] Kindeskog, G., Pettersson, G., 2016. Ambient energy harvesting: A feasibility study and design of test circuits.

[24] Lal, D.K., Dash, B.B., Akella, A.K., 2011. Optimization of PV/wind/micro-hydro/diesel hybrid power system in HOMER for the study area. Int. J. Electr. Eng. Inform. 3, pp 307–325.

[25] Leschin, S., Cass, R., Mohammadi, F., 2007. Self-powered portable electronic device.

[26] Liu, J., 2010. Multifunctional portable energy storage device.

[27] Omer, A.M., 2008. Energy, environment and sustainable development. Renew. Sustain. Energy Rev. 12, 2265–2300.

[28] Pires, V.F., Romero-Cadaval, E., Vinnikov, D., Roasto, I., Martins, J.F., 2014. Power converter interfaces for electrochemical energy storage systems–A review. Energy Convers. Manag. 86, 453–475.
[29] Pitao Jr, R.L., Abing, E.N., Araneta, S.E., Dionson, C.J.C., Gagarino, J.G., Famor, R.M., 2019. Design and construction of a 20 000 Mah wind power bank.
[30] de Rhodes, M.F.M., 2016. Oscillating driving circuit for a wireless power transfer system (PhD Thesis). Instituto Superior de Engenharia de Lisboa.
[31] de Ribeiro, L.A.S., Saavedra, O.R., Lima, S.L., de Matos, J.G., Bonan, G., 2012. Making isolated renewable energy systems more reliable. Renew. Energy 45, 221–231.
[32] Smith, M.G., DeLuca, M.J., Keane, J.A., Geris, R.A., Dill, S.L., Chen, H.Y.-T., Eaton, E.T., Bos, J.C., Veselic, D., 2013. Apparatus, and associated method, for providing charging energy to recharge a portable power supply.
[33] Suen, C.Y.B., Chan, K.T., Hung, T.K., Lee, C.C., 2017. Remote monitoring on capacity of portable power bank in testing laboratories, in: IECON 2017-43rd Annual Conference of the IEEE Industrial Electronics Society. IEEE, pp. 4734–4739.
[34] Tamai, G., Aldrich III, W.L., 2001. System for battery module balancing via variable voltage DC-DC converter in a hybrid-electric powertrain.
[35] Ter-Gazarian, A.G., 1994. Energy storage for power systems. IET.
[36] Wu, D., 2010. Universal battery module and controller therefor.
[37] Yilmaz, M., Krein, P.T., 2012. Review of battery charger topologies, charging power levels, and infrastructure for plug-in electric and hybrid vehicles. IEEE Trans. Power Electron. 28, 2151–2169.

Chapter 4

Maximum Power Point Tracking of PV System using ANN Algorithm

Teoh Chun Kuey, *Ramani Kannan, Rajvinder Singh, Devamurugan, and Wan Muhammad Bahrein

Department of Electrical and Electronics Engineering, Universiti Teknologi Petronas, Malaysia
Corresponding author: ramani.kannan@utp.edu.my

Abstract

Electricity has become a necessary need in Malaysia as in other countries. Energy usage has increased over the years, and electricity prices have increased due to the increase in energy demand. Therefore, people need to pay more than before to access this facility provided by Tenaga Nasional Berhad (TNB) in Malaysia. Due to the increase in prices, people have started to look for an alternative solution to fulfill the electricity demand. Solar energy is a suitable solution for this problem, as it is also a renewable energy source. Malaysia has a sunny and monsoon season which is the peak time for the sunlight to be in contact with the earth's surface area. Therefore, solar radiation is an important factor in providing an efficient amount of expected output. In this chapter, a method has been proposed using an artificial neural network (ANN) to track the maximum power point of the solar cell at any irradiance level to improve the performance of a photovoltaic (PV) system. A multilayer feed-forward perceptron type NN has been used for MPPT control, and a back-propagation algorithm is being used for training purposes. The proposed method can be utilized to optimize the performance of the photovoltaic system.

4.1 Introduction

The monthly bill for electricity has increased over the years. Therefore, an alternative solution was introduced to ease this problem for powering up the household application so that the amount to pay for electricity bills will be lesser. In order to achieve that, an investment should be made in solar energy as it is identified as a suitable solution for this problem. Therefore, solar panels are required to power up a few electrical appliances in a house. To ensure that the amount of power produced from PV cells is maximized, a maximum power point tracking point (MPPT) algorithm is required. In this study, we designed an MPPT algorithm based on a neural network trained using a back-propagation algorithm.

The main objectives of this study are as follows:

a) To understand the basic concept of artificial neural networks and their application.

b) To develop a model-based design for a solar PV system that supplies an AC load. The design should convert the dc voltage produced from the solar panel to 230 V_{ac} to power up the household appliances.

c) To apply an artificial neural network algorithm for the solar PV system.

4.2 Related Literature Review

There are several methods when it comes to choosing the ANN methods. The ANN methods can be categorized under the hill-climbing method. This technique is sub-categorized under three different types: perturb & observe algorithm (P&O), modified adaptive P&O method, and incremental conductance algorithm (INC). Currently, the MPPT algorithm is mainly developed through continuous optimization of PV system mathematics and control models. There is no need to adjust other parameters for the PV system as the direct control of the converter duty cycle. This concludes that the MPPT control structure has good control performance when the external environment is stable. However, disturbance to the duty cycle must be applied by the algorithm to determine the maximum power point (MPP). Thus, when the external environment changes, the working point may deviate from the correct tracking trajectory, which could cause miscalculation. The results of this would be the tracking accuracy of the system and response speed which can cause power loss.

The conventional incremental conductance algorithm detects the slope of the P–V curve through MPP tracking. It uses instantaneous conductance I/V

and incremental conductance dI/dV for MPPT. In reference to [1], depending on the relationship between the two values, the location of the operating point of the PV module in the P–V curve can be determined; for example, $\frac{d_i}{dv} = -\frac{1}{v}$ indicates that the module operates at MPP, whereas $\frac{d_i}{d_v} > -\frac{1}{v}$ and $\frac{d_i}{d_V} < -\frac{1}{V}$ indicate that the PV module operates at the left and right sides of the MPP in the P–V curve.

The efficiencies of P&O and INC algorithms are 96.5% and 98.2%, respectively [3]. The average increase in energy extraction is 16%–43% using conventional hill climbing MPPT [4]. The hill-climbing-based algorithms oscillate around the MPP in slow varying atmospheric conditions. The hill-climbing-based algorithms are only suitable in rapidly changing atmospheric conditions to reduce losses due to oscillations, and the constant voltage method is fast and sufficient for constant conditions. The two-mode control algorithm combines these two algorithms by using the incremental conductance method for more than 30% normalized solar radiation and the constant voltage method for less than 30% normalized radiation [5].

The study will focus on designing a photovoltaic (PV) system and the design of an MPPT system based on the feed-forward neural network algorithm. The system will be designed in the MATLAB Simulink. Also, the system will be simulated to analyze the characterization of the solar PV. The simulation is also done to investigate the efficiency of the system as it should produce a final output of 230 V_{ac}. The power needed will be calculated based on the electrical appliances that will be needed and then the number of solar panels required will also be calculated. The maximum power point tracking point (MPPT) algorithm is designed to maximize the power produced from the solar PV. A boost converter will be designed to achieve a higher output voltage which will be fed into the inverter. An inverter is also designed to change the output voltage of the boost converter (DC voltage) to 230 V_{ac} voltage that will be used for household appliances. Therefore, there will be a total of three different design parts, which are categorized as follows:

a) Modeling of the solar PV panel

b) Designing MPPT algorithm based on feed-forward neural network

c) Modeling and simulation of the DC–DC converter

d) Modeling and simulation of the inverter

4.3 Methodology

This section gives an introduction to back-propagation and how it was implemented into the solar system. The Levenberg–Marquardt back-propagation algorithm that is discussed can be obtained from Neural Network Toolbox in MATLAB.

4.3.1 Back-propagation algorithm

The back-propagation (supervised learning) is used to train multilayer feed-forward networks in the artificial neural network. Gradient descent with respect to weight is computed using this learning algorithm (back-propagation) in an artificial neural network. Using back-propagation in an artificial neural network, the weights for each neuron are being updated from output to input (in the backward direction rather than forward). Back-propagation takes place after each feed-forward passes through a network to adjust the model's parameter based on the change in weight and biases.

4.3.2 Levenberg–Marquardt back-propagation algorithm

The Levenberg–Marquardt back-propagation algorithm is a simple but robust method to approximate a function by providing a numerical solution to the problem of minimizing non-linear function. The Levenberg–Marquardt back-propagation algorithm mixes the Gauss–Newton algorithm with the steepest descent method. It is robust as it can converge well even when the error surface is more complex than the quadratic situation. This Levenberg–Marquardt algorithm around the area performs a combined training process with complex curvature. For the algorithm to make a quadratic approximation, it needs to obtain a proper local curvature and the

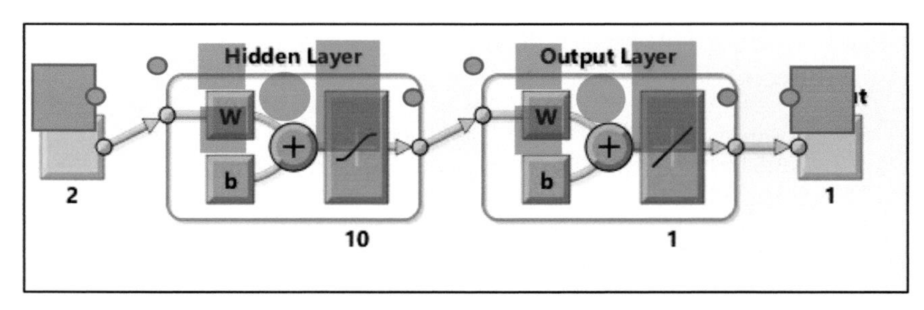

Figure 4.1 ANN MPPT Simulink block.

Levenberg–Marquardt switches can obtain it to the steepest descent algorithm. Then the convergence process significantly sped up once it switched to the Gauss–Newton algorithm. For this ANN MPPT algorithm, we will have two inputs, which are irradiance and temperature. Our targeted output is the duty cycle of the boost converter. The hidden layer will have 10 neurons, which is the default value of MATLAB as shown in Figure 4.1. The number of neurons will be modified later on if the performance of the ANN is not satisfied. For the hidden layer, the log-sigmoid transfer function is used due to differentiable, while for the output layer, the linear transfer function is used to have an output range from 0 to 1. The learning rate of the ANN is the default which is 0.001.

4.3.3 Maximum power point tracking method

MPPT is used to extract from the solar panel and transfer the power from the PV module to the electrical appliance (load). The duty cycle is varied with the change in load impedance, and this change in impedance is then matched at the peak power. The temperature of the cell is a factor that affects the maximum power point.

Perturb and Observe Method:

a) The idea of perturb and observe method is to modify the operating voltage and current of the photovoltaic system until the maximum power is obtained.

b) The PV voltage and current are measured and used to calculate the corresponding power.

c) If the change in power is less than or equal to the pre-set value, then the system in steady-state assumption is made.

Perturb and observe method starts with measuring the voltage and current of the solar PV panel. Power is calculated based on the measured voltage and current value. Perturb and observe will repeat the cycle without any modification if there are no power differences between the previous and current cycles as shown in Figure 4.2. Perturb and observe will have two possible conditions if there is a positive difference in power between the previous and the current cycle. The first condition is that the difference of voltage is less than zero, while the second condition is that the difference of voltage is more than zero. Perturb and observe updates the duty cycle value based on these two conditions.

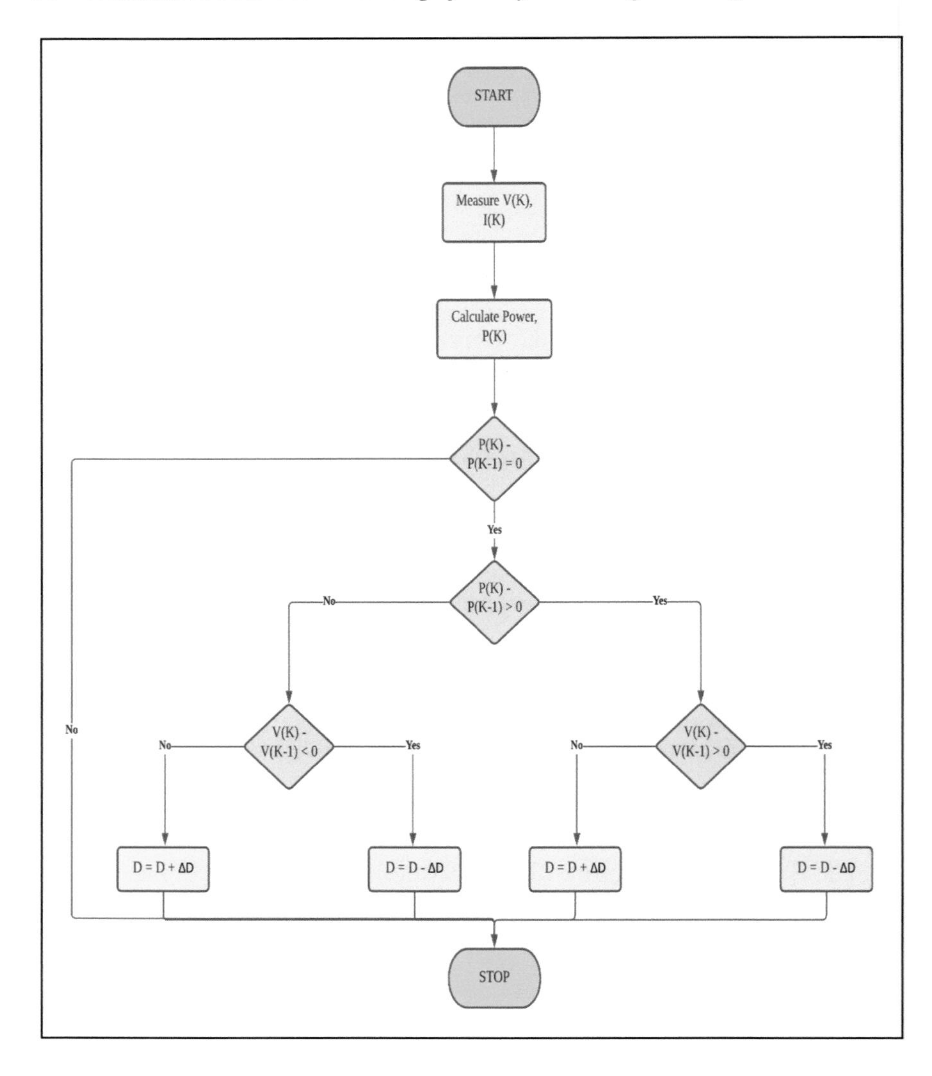

Figure 4.2 Perturb and observe (P&O) flowchart.

4.4 Design Calculation

This section calculates solar PV panel modeling based on three basic electrical loads in a household. A typical power consumption of different house appliances are shown in Table 4.1. The selection of solar PV panel model and size is based on the total power needed by the household. Based on the selected solar PV model, the maximum power condition of the setup is observed under different irradiation and temperature. The solar irradiation graph is designed to mimic the availability of the sun from morning to night.

Table 4.1 Electrical appliance usage in a house.

Appliance	Power consumption (Watts)	Quantity	Hours/ day	Watt-hour/ day
Florescent light	40	2	6	480
Ceiling fan	50	2	8	800
Wi-Fi router	20	1	24	480

4.4.1 Modeling of the solar PV panel

The PV module with the specific capacity *I–V* characteristic curve needs to be used in MATLAB. The selection of the solar panel size is open for the individual to decide, and a detailed analysis of the PV module is expected. Therefore, the usage of electrical appliances needs to be identified.

a) Determine Power Consumption Demands

Total appliance used = (40 W × 6 hours × 2) + (50 W × 8 hours × 2) + (20 W × 24 hours)
= 1760 Wh/day

Backup for at least half day = 1760 × 1.5 = 2220 Wh

Total PV panels energy needed = 2640 × 1.3 (Power loss)
= 3432 Wh/day

b) Size the PV Panel

Total Wp of PV panel capacity needed = 3432/3.4
= 1009 Wp

Number of PV panels needed = 1009/213
= 4.7 modules

This system should be powered by at least five modules of 213 Wp per PV module as shown in Figure 4.3. The actual requirement is nine modules (four parallel and five series to balance the output voltage and current). The system should be powered by 4–8 modules of 213 Wp PV module to obtain more power which can be stored as a backup for a few days.

c) Module Data

Table 4.2 Solar PV parameters.

Parameters	Value
PV module	1Soltech 1STH-215-P
Maximum power (W)	213.15
Open circuit voltage, V_{oc} (V)	36.3
Short-circuit current, I_{sc} (A)	7.84
The voltage at the maximum power point, V_{mp} (V)	29
Current at the maximum power point, Imp (A)	7.35

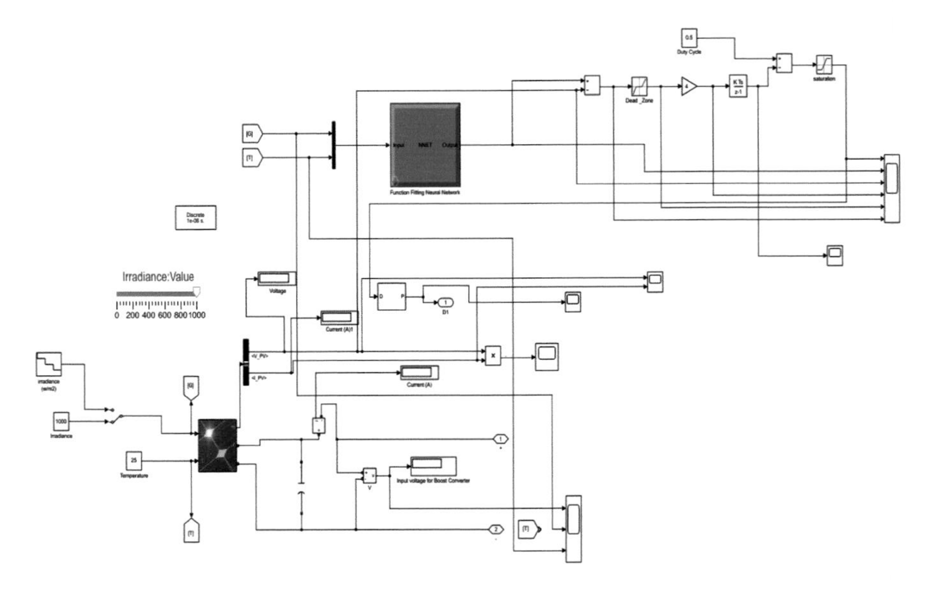

Figure 4.3 PV panel Simulink circuit.

d) Circuit Specifications

Table 4.3 Specifications of the circuit.

Parameters	Value
Irradiance (W/m$_2$)	1000
Temperature (ºC)	25
Stop time (s)	0.02

For this project, the first step was to do the modeling of the solar PV panel. This solar PV array was chosen from the Simulink library. The Simulink library had many types of solar panel arrays. The 1Soltech 1STH-215-P was chosen as stated in Table 4.1. The details of this solar panel can be seen in Tables 4.2 and 4.3. The value of irradiance and temperature has been selected based on Malaysia's average irradiance and temperature. This irradiance and temperature will be varied later on to see how it affects the output. Before confirming the type of PV module that wants to be utilized, the *I–V* and *P–V* graphs were analyzed. Following are the plots as shown in Figure 4.4.

By analyzing these curves, we can see the characteristics of the solar panel. For example, measuring the relationship between current and voltage while varying the electrical load connected to the PV cell or module from open circuit to short circuit produces a characteristic current vs. voltage (*I–V*) curve as shown in Figure 4.1. The points where the curve meets the current

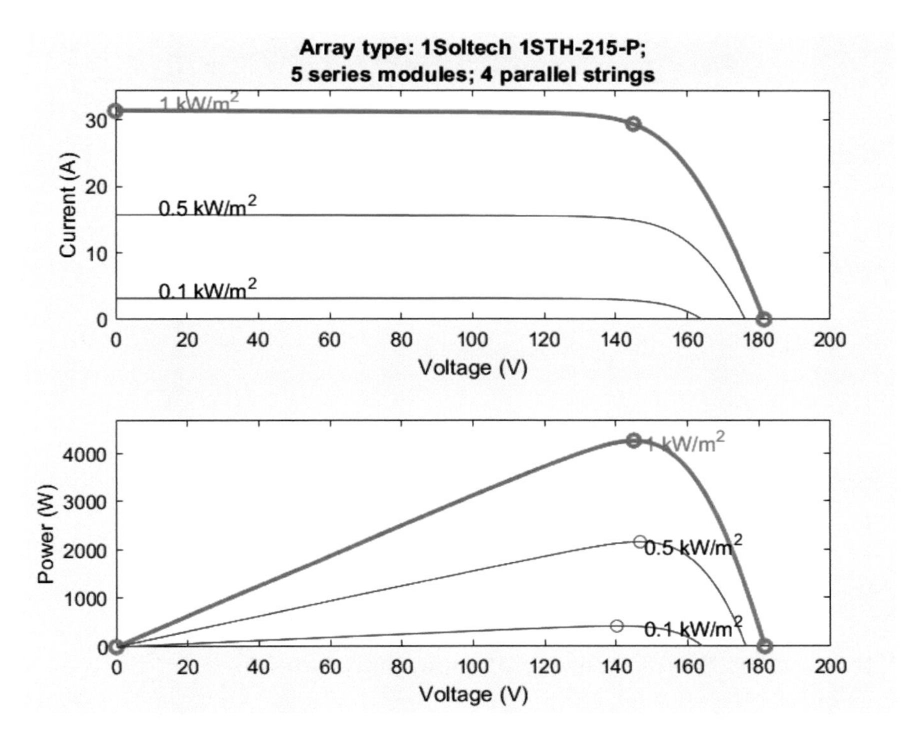

Figure 4.4 *I–V* and *P–V* characteristics of the PV module.

and voltage axes are the short circuit current I_{sc} and open-circuit voltage V_{oc}, respectively. The DC-link capacitor has been added to stabilize the output of the panel. The value of the capacitor has been chosen based on trial and error. The value that was obtained after the trial and error is 1.0 mF.

e) Irradiation

The solar irradiation graph obtained for this experiment is represented below as Figure 4.5.

The *Y*-axis represents the irradiance magnitude, and the *X*-axis represents the hour of a day on the least number of hours of sunlight received. This means the plot above only represents a day with 4 hours of sunlight, assuming it is going to be a rainy day or cold weather blocking the sunlight. It can be observed that the irradiation level reaches up to 1050 W/m^2 for roughly 1 hour per day, assuming the worst weather factor. Therefore, it is required to obtain the expected voltage for an average of 1 hour of charging time. At this 1 hour, the solar PV will operate at the maximum power with maximum efficiency.

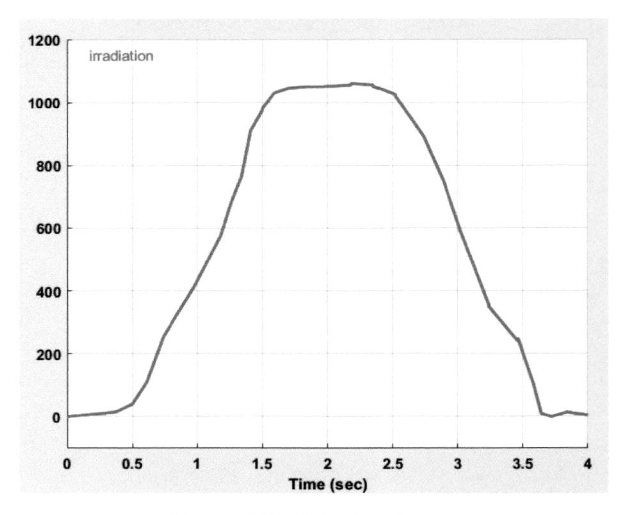

Figure 4.5 Solar irradiation.

4.4.2 Modeling of the DC–DC converter

A circuit of the DC-DC converter is shown in Figure 4.6.

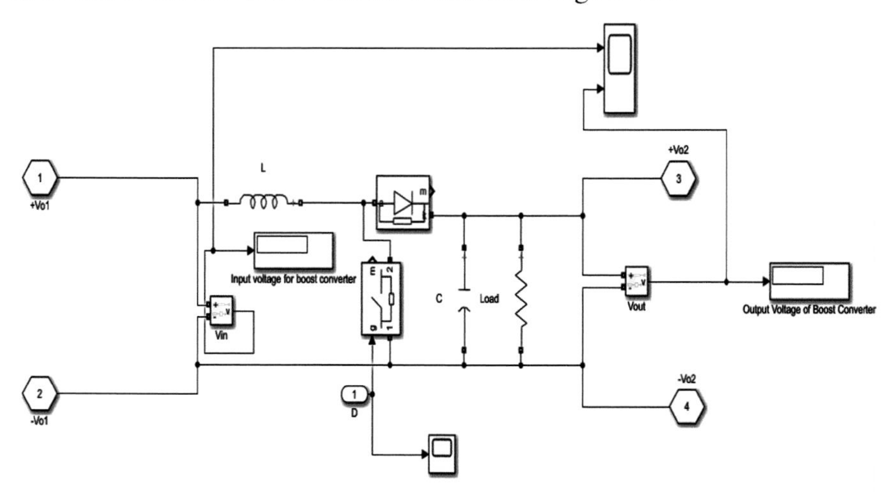

Figure 4.6 Circuit of the DC–DC converter.

Table 4.4 Calculated parameters of the DC–DC converter.

Parameters	Value
Supply DC voltage, V_{in} (V)	145
Output DC voltage, V_o (V)	325

Parameters	Value
Resistor (load), R (Ω)	25
Switching frequency, f_s (Hz)	20 k
Output voltage ripple factor (%)	2
Duty ratio, D	0.554
Critical Inductance, L_{crit} (H)	76.86 μH
Inductor, L (H)	68.874 mH
Capacitor, C (F)	5.50 mF

Table 4.4 shows the calculated values (theoretical) if the DC–DC converter, and, in this case, a boost converter is used. Following are the calculation steps as shown in Figure 4.7.

$V_{in} = 145V \quad V_o = 325V \quad R = 25\Omega \quad f_s = 20khz \quad output\ ripple \leq 2\%$

$$\frac{V_o}{V_{in}} = \frac{1}{1-D}$$

$$\frac{325}{145} = \frac{1}{1-D}$$

$$D = 0.554$$

$$L_{crit} = \frac{DTR}{2} \times (1-D)^2$$

$$L_{crit} = \frac{0.554 \times 25 \times \frac{1}{20k}}{2} \times (1-0.554)^2$$

$$L_{crit} = 6.887 \times 10^{-5}\ H$$

In order for the circuit to be in continuous conduction mode (CCM), $L \gg L_{crit}$

$$L = 6.887 \times 10^{-6} \times 100$$

$$= 6.887\ mH$$

$$\frac{\Delta V_o}{V_o} = 0.02$$

$$0.02 = \frac{D}{CRf_s}$$

$$C = \frac{0.554}{0.02 \times 25 \times 20 \times 10^3}$$

$$= 5.54 \times 10^{-5}\ F$$

$$= 55.4\ uF$$

Figure 4.7 Calculation for DC–DC converter.

4.4.3 Modeling of the inverter

A typical circuit of inverter is shown in Figure 4.8.

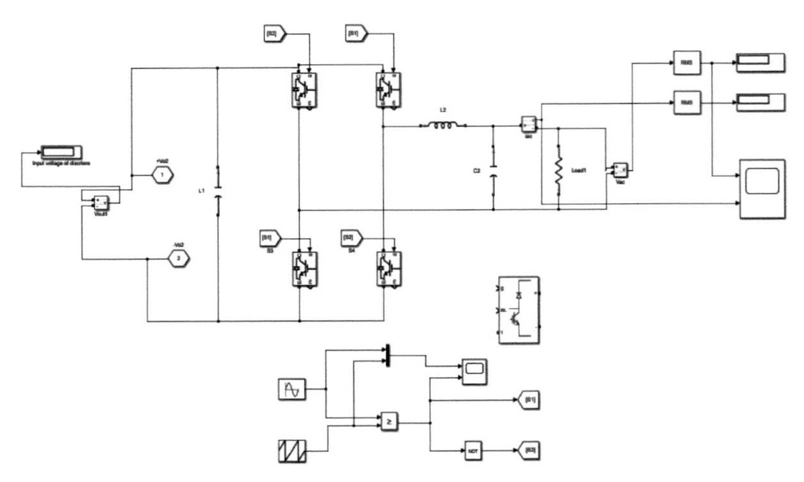

Figure 4.8 Circuit of inverter.

Table 4.5 Parameters for the three-level H-bridge.

Parameters	Value
Supply DC voltage, V_{dc} (V)	325
Resistor (load), R (Ω)	118
Capacitor, C (μF)	28
Inductor, L (mH)	366
Sinusoidal frequency, f_r (reference signal) (Hz)	50
Sawtooth frequency, f_c (carrier signal) (kHz)	5

For this project, a full-bridge inverter with an SPWM circuit. This is also known as a three-level H-bridge. Table 4.5 shows the parameters for this inverter circuit. The function of an inverter circuit is to convert direct current (DC) to alternating current (AC). One of the main differences between a half-bridge and a full-bridge is that the output voltage of the full-bridge inverter is equal to the power supply voltage ($V_s = V_{dc}$).

In contrast, the output voltage for the half-bridge inverter is equal to one half of the power supply voltage ($V_s = V_{dc}/2$). Since the output power is proportional to the square of the voltage, the output power of the full-bridge will be four times that of the half-bridge. Therefore, the full-bridge circuit has been chosen. This inverter is with SPWM. SPWM stands for sinusoidal pulse width modulation, and it is better than a standard full-bridge inverter as the switching frequency is equal to that of the carrier wave. Thus, one of the benefits is having a constant switching frequency. At the same time, this technique improves harmonic characteristics, reduces the switching power devices, and decreases the switching losses.

4.4.4 Final design of solar PV system

A circuit design of proposed solar PV system is shown in Figure 4.9.

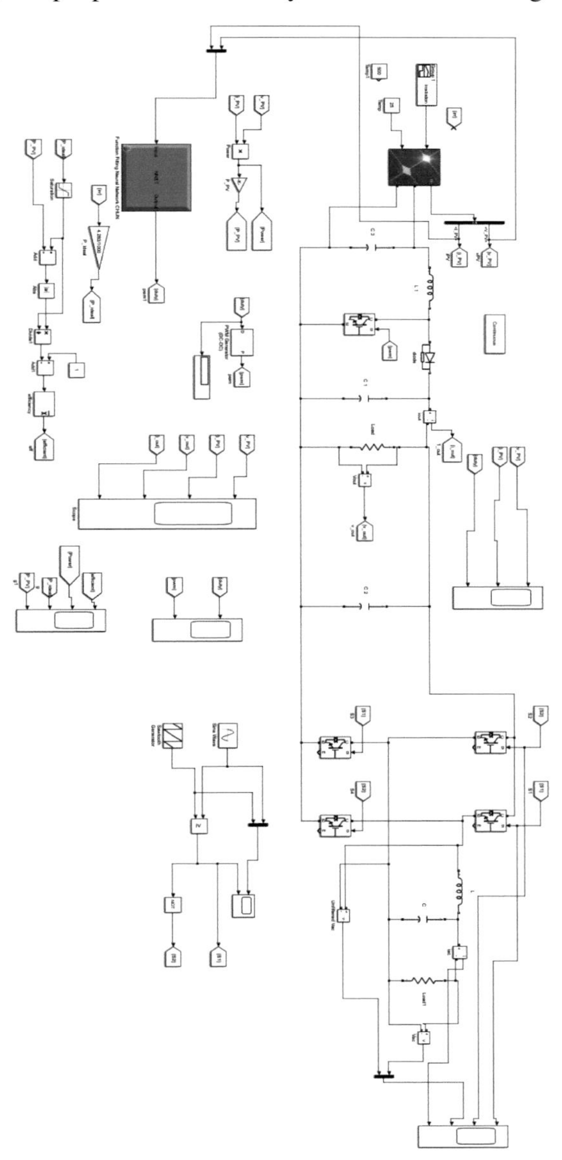

Figure 4.9 Solar PV circuit.

4.5 Result Analysis and Discussion

4.5.1 ANN-based maximum power point tracker (MPPT)

Before ANN MPPT is implemented into our solar system, the ANN must be trained first. During the training of the data, the initial weight is picked randomly. The weight will be adjusted for every iteration until the performance mean squared error is fulfilled. A second training is conducted to ensure our ANN MPPT is trained completely. The second training has only 6 iterations compared to 766 iterations in the first training. According to MATLAB Neural Network guidelines, iterations of 6 means the system has reached the optimal mean squared error performance. The training of ANN MPPT is shown in Figures 4.10 and 4.11.

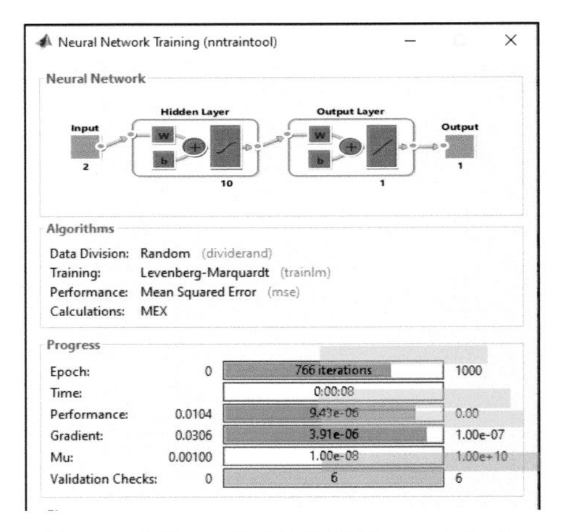

Figure 4.10 ANN MPPT (first training).

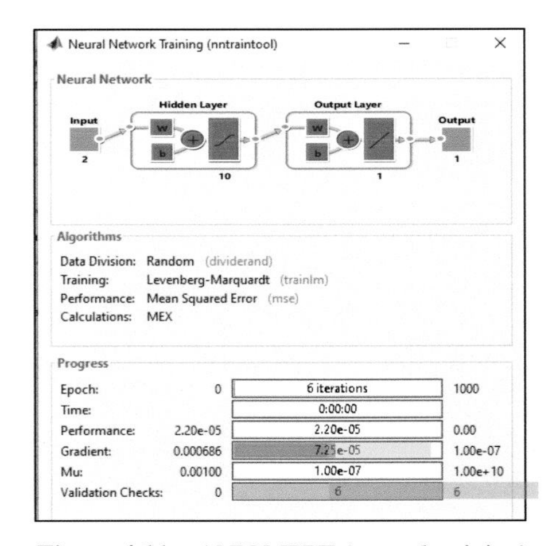

Figure 4.11 ANN MPPT (second training).

The weight of layers are shown in Figures 4.12 and 4.13, whereas the bias of layer 1 is shown in Figures 4.14 and 4.15.

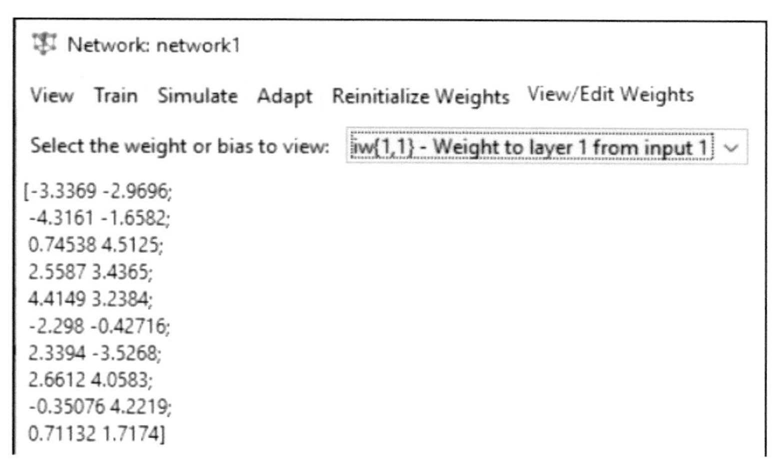

Figure 4.12 Weight to layer 1 from input 1.

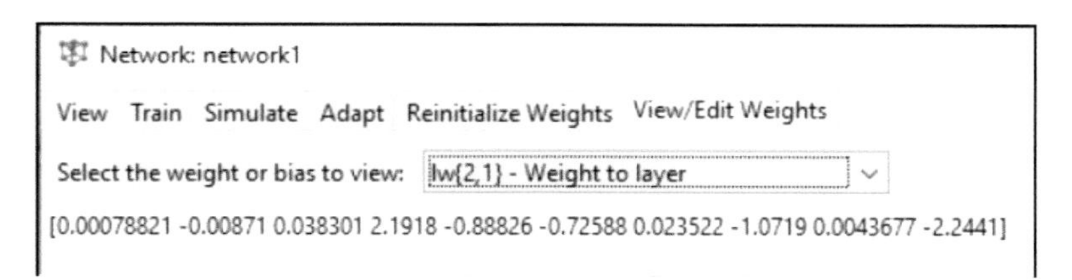

Figure 4.13 Weight to layer.

Network: network1

View Train Simulate Adapt Reinitialize Weights View/Edit Weights

Select the weight or bias to view: b{1} - Bias to layer 1

[4.3615;
3.1772;
-0.95283;
1.4931;
1.3228;
-1.6712;
2.7158;
2.3085;
-3.5843;
1.5065]

Figure 4.14 Bias to layer 1.

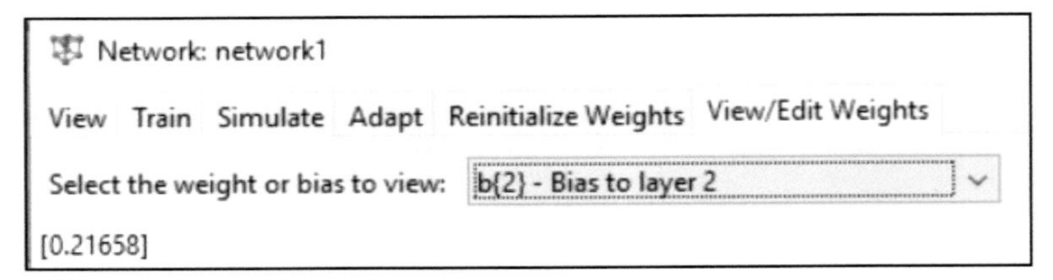

Figure 4.15 Bias to layer 2.

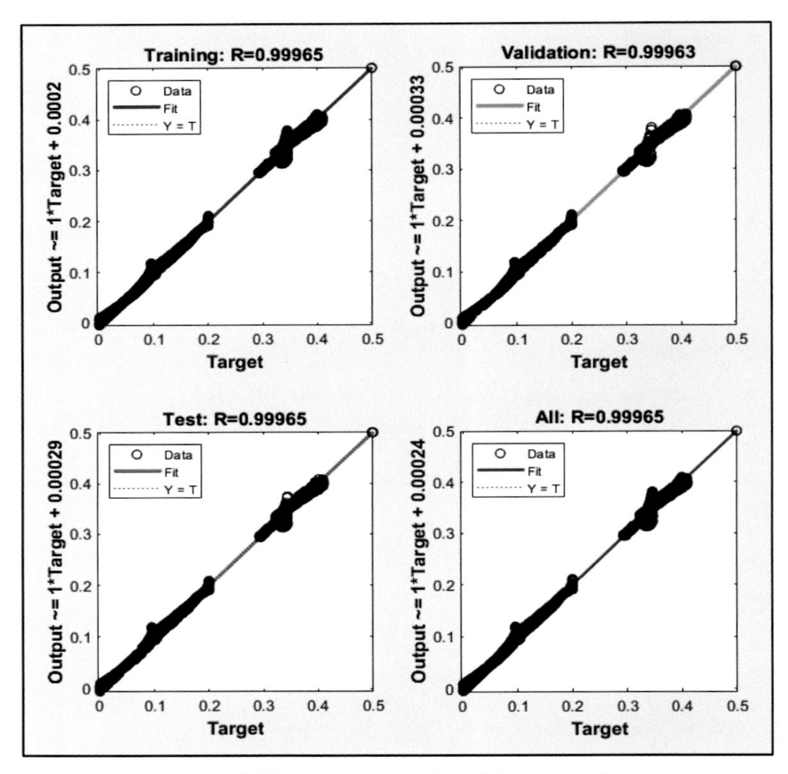

Figure 4.16 Neural network training regression.

We can see how the neural network is learning by observing the weight. The neural network is fed forward and back-propagation various times to readjust the weight value to ensure minimal error between the targeted and simulation output. The neural network will keep on training, and, for our case, the algorithm has fine-tuned its weight value up to 766 times due to the sizes of the training data. Once the neural network has finished training, we can observe its performance in Figure 4.16. All of the targeted data fit nicely into the fitting line, and the regression value is approximately 1.

4.5.2 Output observation

The output observed at Solar PV and MPPT is shown in Figures 4.17 and 4.18, respectively.

a) Output Observed at the Solar PV

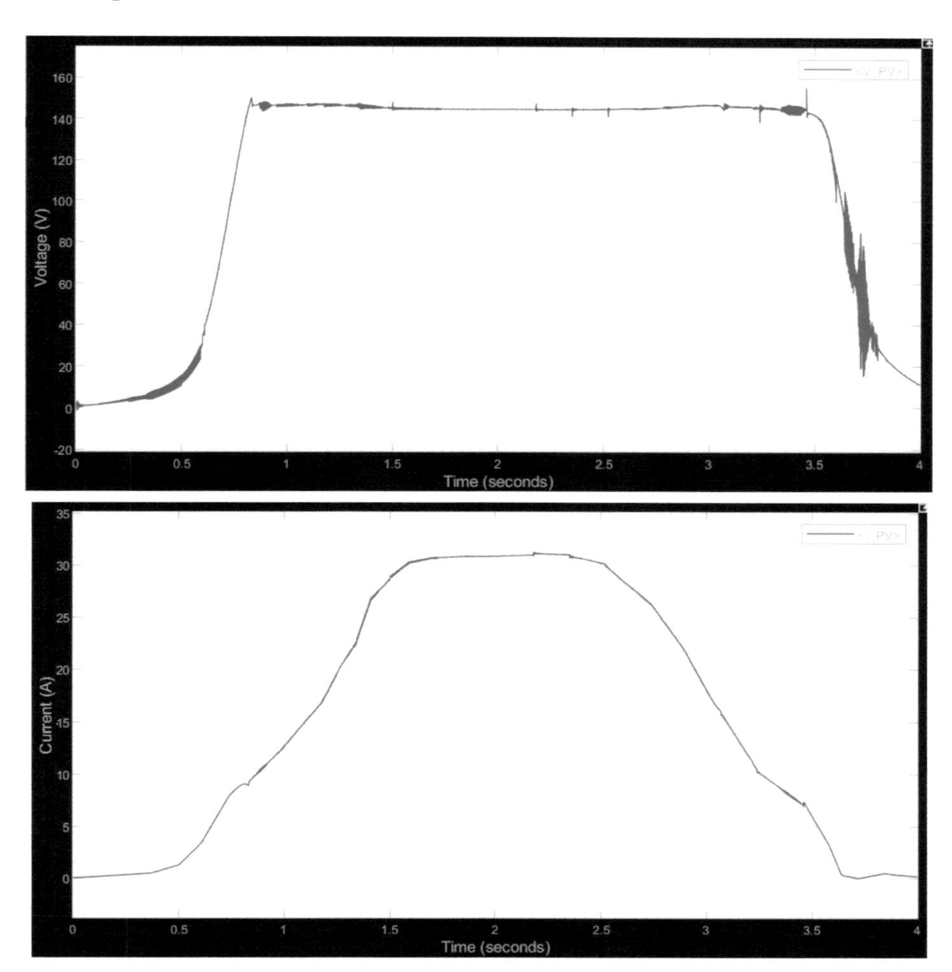

Figure 4.17 Output current, output voltage at solar PV.

Overall observed waveform from $T = 0$ to $T = 4$. Therefore, the irradiance will change according to the time, same as the voltage and current produced.

b) Output Observed at Maximum Power Point Tracking (MPPT)

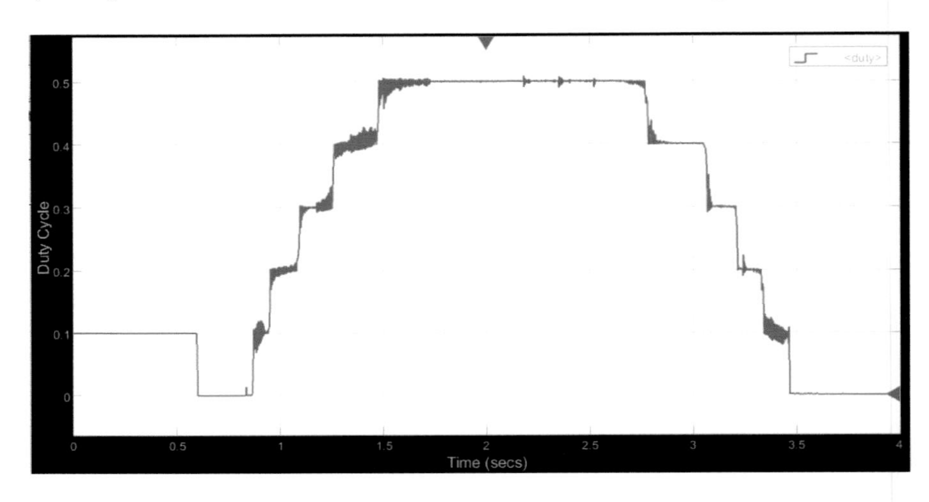

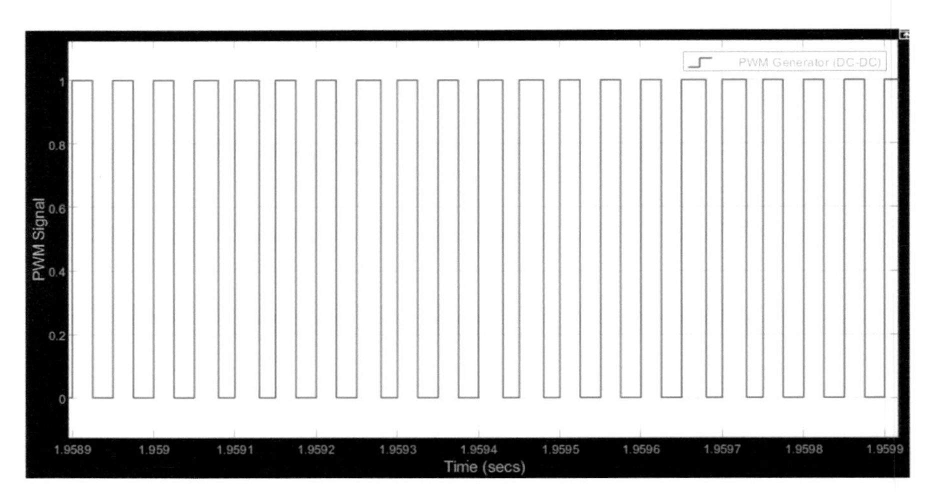

Figure 4.18 Output observed at MPPT.

The obtained graph above shows the changes in the duty cycle due to the MPPT algorithm. At every second (continuous), the MPPT will read the power value and store it as the previous value as it reads a new value. Then it will compare the new value with the previous value. If the new power value exceeds the previous value, the duty cycle will increase by adding the step size value until it reaches the maximum power point. At seconds after the maximum power value point is achieved, the new value will be lesser than the previous value, and, this time, the duty cycle will decrease with each step. The new power value is lesser than the previous value. As for the PWM, it will be at 1 (turn ON) as long as the duty cycle does not reach zero. If the duty cycle reaches zero, then the PWM will also become 0 (turn OFF).

c) Output Current and Voltage for DC–DC Boost Converter

The observations of output current and voltage of DC-DC boost converter is shown in Figures 4.19, 4.20 and 4.21.

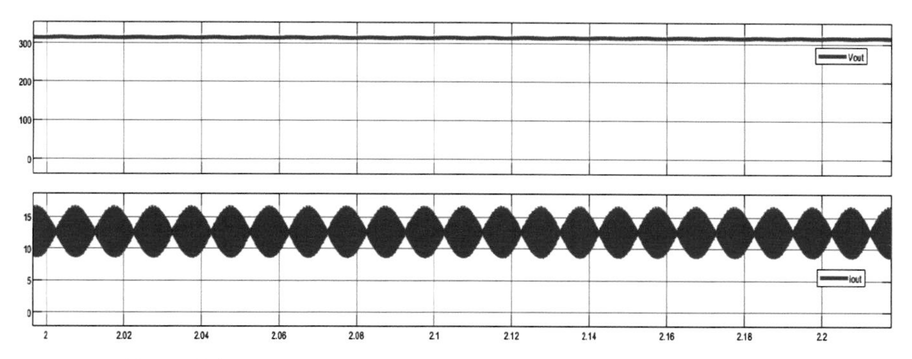

Figure 4.19 Output current and voltage for DC–DC boost converter.

Trace Selection

Vout

Signal Statistics

	Value	Time
Max	3.155e+02	2.174
Min	3.136e+02	1.939
Peak to Peak	1.867e+00	
Mean	3.146e+02	
Median	3.146e+02	
RMS	3.146e+02	

Figure 4.20 V_{out} from DC–DC boost converter.

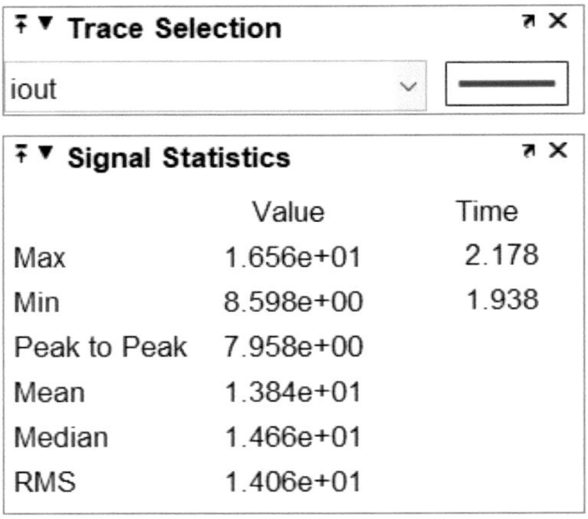

Trace Selection

iout

Signal Statistics

	Value	Time
Max	1.656e+01	2.178
Min	8.598e+00	1.938
Peak to Peak	7.958e+00	
Mean	1.384e+01	
Median	1.466e+01	
RMS	1.406e+01	

Figure 4.21 I_{out} from DC–DC boost converter.

There is a voltage drop at the output of the boost converter after connecting it to a full-bridge inverter as listed in Table 4.6.

Table 4.6 DC–DC boost converter output.

Parameter	V_{out} Avg (V)	I_{out} Avg (A)
Value	314.6	13.86

The calculated parameters are similar to the values obtained from the simulation. The variance is because the internal resistance (losses) is considered in the simulation. This has caused a voltage drop (small value); therefore, the value obtained from the simulation tends to be slightly smaller than the calculated value (with no loss assumption). The critical inductor value is (L = 68.9 μH). In this experiment, the L value is greater than the L (critical). Therefore, the output voltage will be in the range of the calculated average voltage (V_{avg} = 325). The obtained V_{avg} is 314.6 V.

d) Load Variation for Boost Converter

Given: f_s = 20 kHz, D = 0.554, L_{crit} = 68.87 μH, and C = 55.4 μF.

It can be observed that the voltage and current values changes as the load value is changed as listed in Table 4.7. It is safe to say that the load value 25 is used for this experiment to get the expected values. Observing this variation deeper, we know that load is not the only manipulated variable that changes in a boost converter. As we can see, the other parameters are kept the same while varying the load. The output obtained from this action is not what we need. Capacitance and inductance values are very dependent on the load value as it changes accordingly with the change in load value.

Table 4.7 Load variation.

Load	V_{out}, V	I_{out}, I
10	205.8	21.39
25	314.6	13.86
40	384.9	11.15

e) Output Observed at H-Bridge Inverter

The various observation of H-bridge inverter is shown Figures 4.22, 4.23, 4.24, 4.25, and 4.26.

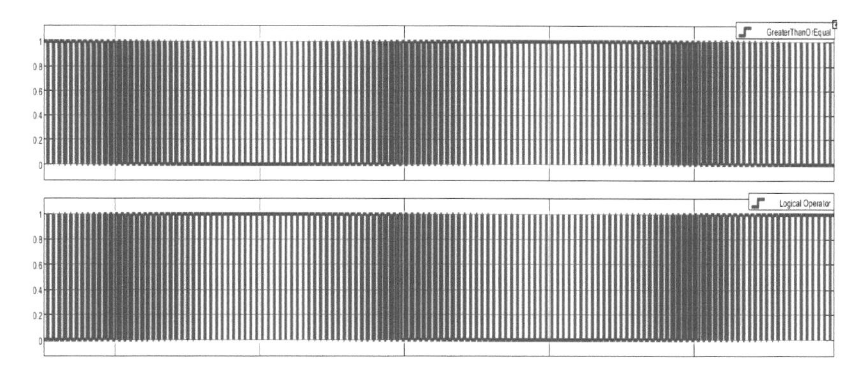

Figure 4.22 Logical operator at H-bridge inverter.

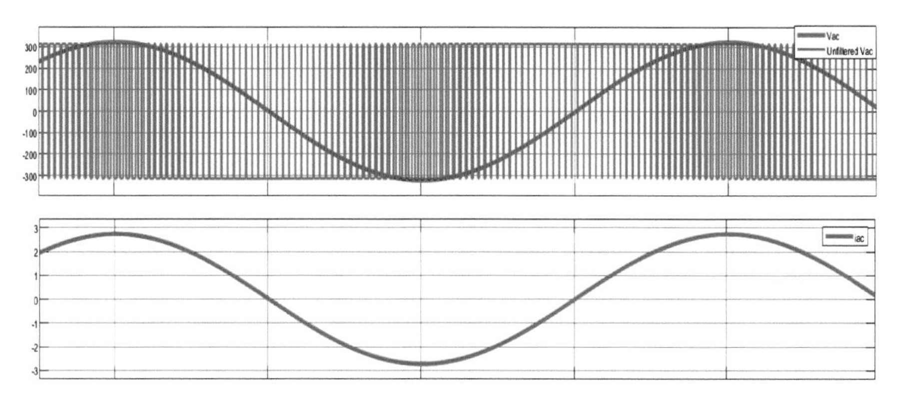

Figure 4.23 Output V_{ac}, unfiltered V_{ac}, and I_{ac} of H-bridge inverter.

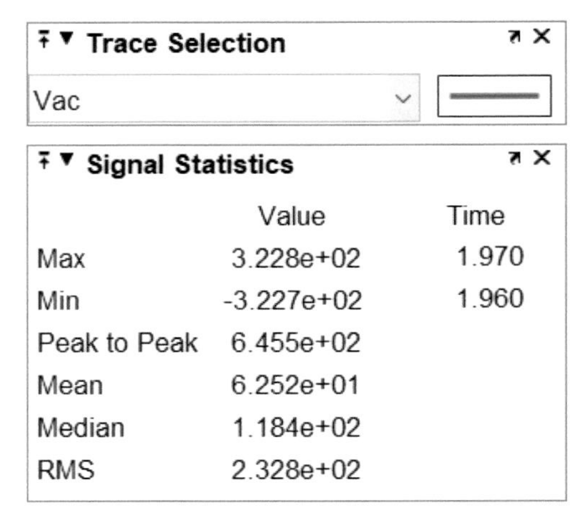

Figure 4.24 Output V_{ac} of H-bridge inverter.

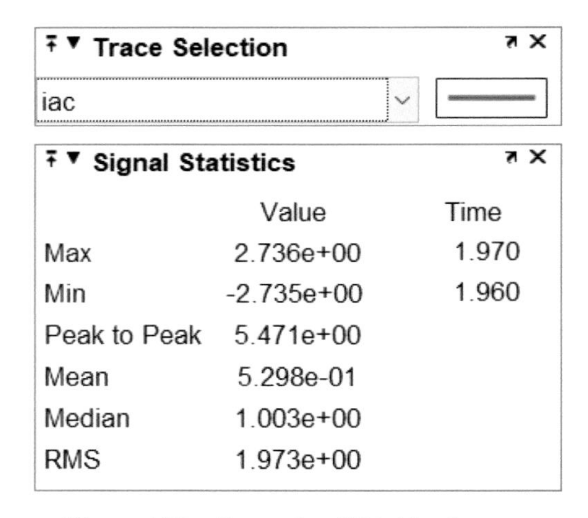

Figure 4.25 Output I_{ac} of H-bridge inverter.

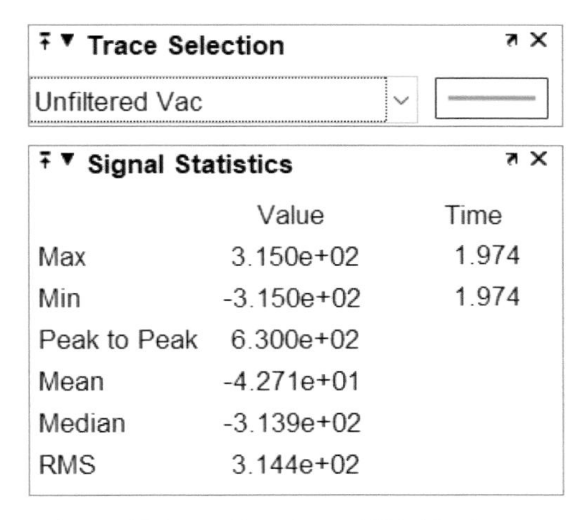

Figure 4.26 Unfiltered V_{ac} of H-bridge inverter.

Table 4.8 Output of H-bridge inverter.

Parameter	**V_{ac} (V)**	**I_{ac} (A)**	**Unfiltered V_{ac}**
Value	230.6	1.95	237

The $V_{rm} = (V_{peak} \times 0.707)$ voltage obtained is 230 V_{ac}, the nominal AC voltage used to power up household appliances. The output of H-bridge inverter is shown in Table 4.8. Therefore, it is required to design a system that does produce the expected voltage. Therefore, the AC voltage designed will have a peak-to-peak voltage of 650 V. It can be observed from the graph obtained above that the desired output is achieved from the designed system. This inverter consisting of two basic circuits can produce an output voltage twice as much as that of the half-bridge inverter using the same DC voltage. In the single-phase full-bridge inverter, the load voltage is the difference between two pole voltages:

$$V_o = V_H - V_L.$$

To produce the effective load voltage in this inverter, the phase difference between two pole voltages needs to be 180°. This voltage is double the voltage of the half-bridge inverter under an equal DC voltage.

4.6 Conclusion

This project-based experiment was conducted through MATLAB Simulink to develop a model-based design for a solar PV system that supplies an AC load. The design should convert the DC voltage produced from the solar panel to 230 V_{ac} to power up the household appliances. Through this project, we were able to apply the knowledge in designing boost converters and H-bridge inverters based on the desired output and also implement artificial neural network into an application related to our electrical and electronics field. We were able to understand the basic concepts of artificial neural networks. We have also applied artificial neural network algorithms for the solar PV system by using MATLAB. Lastly, we also have analyzed the solutions for ANN control algorithms using a solar PV system with power electronics converters. ANN MPPT is widely used in the market for its quick response time compared to conventional MPPT. In conclusion, artificial neural networks can be implemented in various fields, and many applications can be derived from this technology.

References

[1] L. Shang, H. Guo, and W. Zhu, "An improved MPPT control strategy based on incremental conductance algorithm," Prot. Control Mod. Power Syst., vol. 5, no. 14, 2020, pp. 1–8.

[2] A. Anzalchi and A. Sarwat, "Artificial neural network-based Duty Cycle estimation for maximum Power Point tracking in Photovoltaic systems, *SoutheastCon* 2015, pp. 1–5, DOI: 10.1109/SECON.2015.7132988

[3] D.P. Hohm and M.E. Ropp, "Comparative study of maximum power point tracking algorithms using an experimental, programmable, maximum power point tracking test bed," Conference Record of the Twenty-Eighth IEEE Photovoltaic Specialists Conference, pp. 1699–1702, 2000.

[4] J.H.R. Enslin, "Power point tracking: A cost saving necessity in solar energy systems," 16th Annual Conference of IEEE Industrial Electronics Society, vol. 2, pp. 1073–1077, 1990.

[5] G.J. Yu, Y.S. Jung, J.Y. Choi, and G.S. Kim, "A novel two-mode MPPT control algorithm based on comparative study of existing algorithms," Solar Energy, vol. 76, 2004, pp. 455–463.

[6] S.B. Riffat and X. Ma, "Thermoelectrics: A review of present and potential applications," Applied Thermal Eng., vol. 23, no. 8, pp. 913–935, Jun. 2003.

[7] T. Esram and P.L. Chapman, "Comparison of photovoltaic array maximum power point tracking techniques," IEEE Trans. Energy Conversion, vol. 22, no. 2, pp. 439-449, Jun. 2007.
[8] R.-Y. Kim and J.-S. Lai, "A seamless mode transfer maximum power point tracking controller for thermoelectric generator applications," Conference record of the IEEE Industry Applications Conference, New Orleans, LA, USA, Sep. 2007, pp. 977-984.
[9] H. Nagayoshi, T. Kajikawa, and T. Sugiyama, "Comparison of maximum power point control methods for thermoelectric power generator," International Conference on Thermoelectrics, Long Beach, CA, USA, Aug. 2002, pp. 450–453.
[10] D.P. Hohm and M.E. Ropp, "Comparative study of maximum power point tracking algorithms using an experimental, programmable, maximum power point tracking test bed," IEEE Photovoltaic Specialists Conference, Anchorage, AK, USA, Sep. 2000, pp. 1699–1702.
[11] X. Weidong and W.G. Dunford, "A modified adaptive hill climbing MPPT method for photovoltaic power systems," IEEE Power Electronics Specialists Conference, Aachen, Germany, Jun. 2004, pp. 1957–1963.
[12] H. Chihchiang and S. Chihming, "Study of maximum power tracking techniques and control of DC/DC converters for photovoltaic power system," IEEE Power Electronics Specialists Conference, Fukuoka, Japan, May 1998, pp. 86–93.
[13] R.-J. Wai, W.-H. Wang, and J.-Y. Lin, "Grid-connected photovoltaic generation system with adaptive step-perturbation method and active sun tracking scheme, IEEE Industrial Electronics Conference, Paris, France, Nov. 2006, pp. 224–228.
[14] L. Jae Ho, B. Hyun Su, and C. Bo Hyung, "Advanced incremental conductance MPPT algorithm with a variable step size," EPE Power Electronics and Motion Control Conference, Portoroz, Slovenia, Aug. 2006, pp. 603–607.
[15] L. Bangyin, S. Duan, F. Liu, and P. Xu, "Analysis and improvement of maximum power point tracking algorithm based on incremental conductance method for photovoltaic array," International Power Electronics and Drive Systems Conference, Bangkok, Thailand, Nov. 2007, pp. 637–641.
[16] J. Liang, S.K.K. Ng, G. Kendall, and J.W.M. Cheng, "Load signature study—Part I: Basic concept, structure, and methodology," IEEE Trans. Power Delivery, vol. 25, no. 2, pp. 551–560, Apr. 2010.

Chapter 5

BIM- and GIS-Based Residential Microgrid Modelling: Possibilities, Benefits, and Applications

Jasim Farooq[1], Rupendra Kumar Pachauri[1], and Sreerama Kumar R.[2]

[1]School of Engineering, University of Petroleum & Energy Studies, India
[2]Department of Electrical and Computer Engineering, King Abdulaziz University, Saudi Arabia

Abstract

An increase in the share of solar energy in the grid may unestablish the grid. To overcome the issues of grid instability, specifically in remote areas, the building information modeling (BIM) and geographical information system (GIS) based microgrid planning based on data can be effectively used. The integration of BIM and GIS information systems for microgrid planning is appealing due to its potential benefits, such as it decreases the microgrid planning time and cost. Building BIM- and GIS-based integrated approach for microgrid planning are advantageous over the traditional multi-platform based 2D-CAD modeling process due to increased automation and digitization of the modeling process. This chapter provides consolidated details of the BIM- and GIS-based integrated microgrid pre-construction planning, its applications for regional-level studies, and its advantages over the conventional approach. BIM information system provides building-level electrical data and GIS integrates the building-level data for city-level microgrid planning studies. BIM-based add-in tools is ideal for residential-level bottom-up microgrid planning projects than the GIS-based approach. BIM offers minute level details of the electrical network comparatively easy. By this integrated method, infrastructure data and microgrid optimization techniques are processed simultaneously in a platform which is not feasible in the conventional approach. Further research is required to develop matured applications for

BIM- and GIS-based microgrid planning by combining sophisticated algorithms and modeling standardization to be implemented by government agencies.

5.1 Introduction

BIM is an information system-backed *n*th-dimensional modeling of infrastructure, and it is utilized to improve the visualization, coordination, and efficiency of construction works [1]. By modeling the electrical systems using BIM applications, it is possible to conduct automated code checking and electrical circuit continuity check, design calculations, energy analysis, and quantifications within the BIM platform itself, and, thus, it eases prefabrication work [2]. By utilizing GIS techniques, electrical network modeling, power plant location optimization, big data analytics for electrical-related data normalization, spatial query, and integrated analysis with other disciplines are possible [3]. Integrating BIM and GIS in infrastructural projects is an innovative method and emerging development in current years, from studies to construction practice. BIM electrical models possess project-level geometrical and parametric details of every electrical end load, while GIS can analyze a broad geographic level [4].

By developing BIM-based add-in tools, it is possible to extract and process the required data from the BIM electrical models and power system calculations, quantifications, and optimizations related to microgrid [5, 6]. A residential microgrid planning requires much detailed information related to each and every electrical element and the same detailed modeling cannot be easily performed in GIS. The BIM platform is better suitable for this purpose. After studying BIM and GIS capabilities, the relative capabilities for residential microgrid modeling are given in Table 5.1 [2, 3, 7–10]. In general, the BIM file size for a building is in megabytes. A high level of data transfer between GIS and BIM systems may cause interoperability issues if all the electrical system models are integrated from one platform to another. Only the necessary data for microgrid planning is to be extracted and integrated between BIM and GIS without any data loss. As per the requirements, different levels of data extraction and exchange approaches are possible for microgrid planning, and two such possibilities are given in Figure 5.1 [11].

These discussions show that the BIM platform is suitable for residential microgrid optimization studies. GIS data can be integrated with BIM

Table 5.1 BIM and GIS comparison for a residential microgrid modeling and planning.

Requirements	BIM	GIS
Parametric electrical network modeling	Automatic and easy	Complex
Electrical element modeling	Easy	Complex
3D modeling	Easy and default	Complex
Electrical load details modeling	Easy	Complex
Building/project integrated modeling with all trades	Easy	Complex
Inbuilt tools available for estimation and clash detection between artifacts	More	Less
Add-in tool development	Possible	Possible
Geocoding	Possible	Possible
Weighted overlay analysis with geodata	May be possible	Yes
Regional-level power analysis	May be possible	Yes
Sensor integration in real time	Possible	Possible
Integrated building energy analysis	Sure	No
Occupant behavior influences the energy consumption according to building artifacts organization	Possible	May be

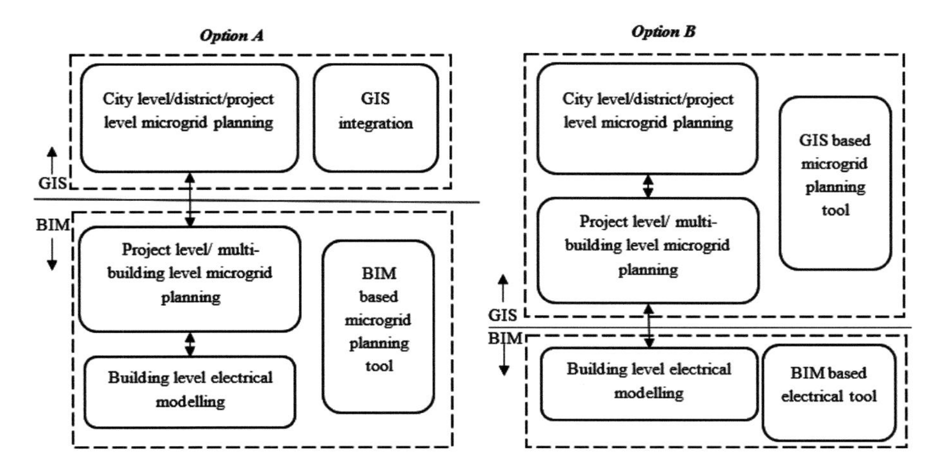

Figure 5.1 BIM- and GIS-based microgrid modeling approaches for microgrid planning.

to conduct site suitability data and gather resource potential. The necessary data from the BIM-based microgrid can be exported to GIS for regional-level analysis. As described in Figure 5.2, the planning process will offer an integrated approach that increases the accuracy of data exchange. Thus, multiple levels of optimizations at local and regional levels can be conducted.

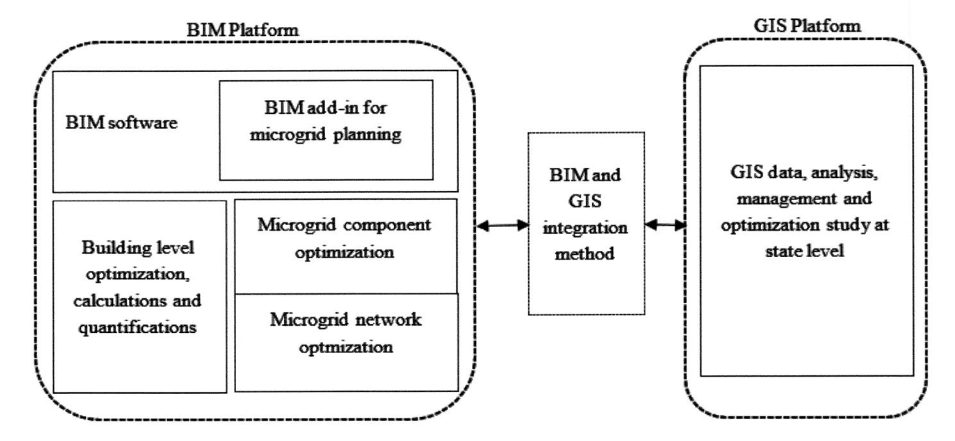

Figure 5.2 BIM- and GIS-based approach for microgrid planning.

5.2 Advantages of BIM- and GIS-Based Microgrid Modeling over Traditional 2D-CAD Approach

In the conventional method, after the microgrid design calculations are done, the results are shown in the drawings with the aid of 2D-CAD software. After the electrical system drafting is done as per client requirements, corresponding electrical measures/materials are physically retrieved from the 2D drawings and are summarized either using Excel-based spreadsheets or in the 2D-CAD software platform itself for complete pricing and the preparation of total electrical load data. BIM-based data gathering is from a single consolidated model where design calculations, quantifications, and analysis can be performed fully under the BIM environment [12].

Because of the absence of arithmetic interlinking between electrical panel board data and corresponding electrical load in the 2D-CAD method, it is essential to manually update all the related data, resulting in more time delays and budgetary losses. In BIM, an information system is used to arithmetically interlink the electrical network along with geometrical details [13].

All building design calculations and drawings must undergo a building code evaluation process in that region. Automated tools are crucial to comply with corresponding building code and regulations, where authenticated codes can be checked automatically with minimal designer involvement. These are ever more desirable in the building construction industry due to sustainability and energy conservation consideration. Compared to 2D-CAD, BIM offers a better platform for automated code checking due to the object-oriented information model-based concept [14].

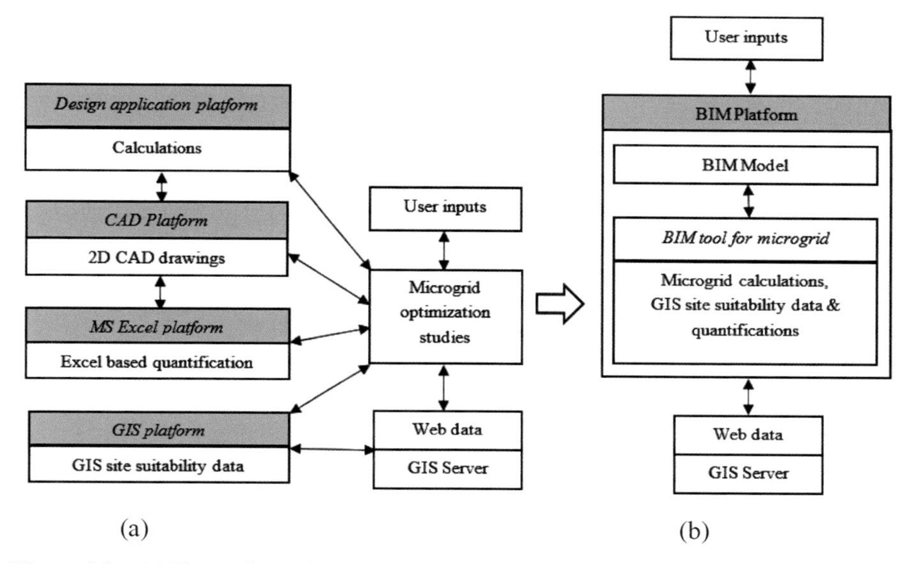

Figure 5.3 (a) The traditional method of the microgrid planning process. (b) BIM- and GIS-based microgrid planning process.

As the BIM-based microgrid data is stored in a coherent model and electric network connection details, data sharing with GIS will be easier than 2D-CAD drawings. As shown in Figure 5.3(a), the conventional microgrid planning process is a multi-platform-based approach where data exchange between platforms is manual. As multiple revisions are normal in an iterative optimization process, the traditional approach is laborious, and the chances of compounding errors are high. By implementing a BIM-based approach given in Figure 5.3(b), a single platform-based approach that can reduce calculation time and increase automation in data exchange is possible. This approach combines infrastructural data and microgrid calculations in a single digitized, sharable platform.

5.3 Data Retrieval from BIM Models using Add-In Tools

Innovations in BIM deliver plenty of opportunities for novel BIM-based tools that can automate the modeling and analysis of construction activities. BIM-based add-in tools can retrieve relevant data from BIM models, and the necessary calculations can be made by processing the gathered data. The following subsections detail the methods for extracting information from BIM models [15, 16].

5.3.1 API-based extraction and processing of information

Using application programming interface (API) based data extraction, available data from BIM models can be extracted and processed to get desired results. By using add-in tools, it is possible to create, edit, add, and retrieve the graphical/family attributes/edit the family parameter of a particular BIM document, such as information for conducting electrical demand estimation according to microgrid optimal planning requirements [17].

5.3.2 API-based extraction and ontology-based processing of information

Suppose the required information is missing from BIM elements, then algorithms can be developed using historic data to generate the necessary values from the parameters of various elements available in BIM. It is possible to generate the necessary values from the parameters of various elements available in BIM. For example, by extracting specifications from BIM elements and using an algorithm to identify the material, compared with the historical database, required information such as material cost can be estimated. The accuracy of this method depends on the precision of the ontology/algorithm utilized for this purpose. For extracting information from a disconnect switch, the specifications described in Table 5.2 are to be extracted from the corresponding BIM elements for estimating the cost of the material. The extracted parameters are matched with the historical cost database and estimated as the nearest possible material cost [18].

Table 5.2 Specifications required for costing a disconnect switch.

Description	Influence on price
Nominal current rating	Material cost directly proportional to the current rating
Nominal voltage rating	Material cost directly proportional to the voltage rating
With fuse?	Fuse
Gasketed?	The additional cost is required for the gasket
Ingress protection (IP) rating	Material costs increase with a high IP rating
Phase	Material cost is directly related to the number of phases
Material	If good material, then the material cost will be more
Manufacturer	If a well-known manufacturer, then the material cost will be more

5.3.3 Data extraction from available analyzing tools

Necessary authorization and programming knowledge are required for developing novel customized add-in tools to extract data from third-party applications or inbuilt BIM tools [19].

5.3.4 Hybrid data extraction

Combining these three methods given in Sections 5.3.1, 5.3.2, and 5.3.3 makes it possible to retrieve information from BIM elements and compare the outcome from different methods for increasing the accuracy of information. For example, load demand estimated by summing up the elementary loads can be authenticated by comparing the results with VA/m^2 according to regional codes using a developed add-in tool. Table 5.3 presents the comparative advantages of the extraction process.

Hybrid data extraction (method-D) is suitable for minimizing the error. As shown in Figure 5.4, data is extracted by method-A, and by using an ontology/algorithm, the extracted information is compared with standard data or third-party applications to increase the accuracy.

5.4 Data Exchange Between GIS and BIM

Both GIS and BIM play distinctive and important roles in assisting the modeling and management of infrastructure projects throughout their lifespan. Modeling paradigms used by BIM and GIS are different; so a smooth data exchange methodology is considered for developing BIM and GIS integrated

Table 5.3 Comparative advantage of BIM-based data extraction process.

Description	Method-A	Method-B	Method-C	Method-D
Chances of error in final results	Less	More	Less	Very less
Complexity of algorithm	Low	High	Low	High
Programming requirement	Less	More	Very less	More
License required from third-party applications	No	No	Maybe	No

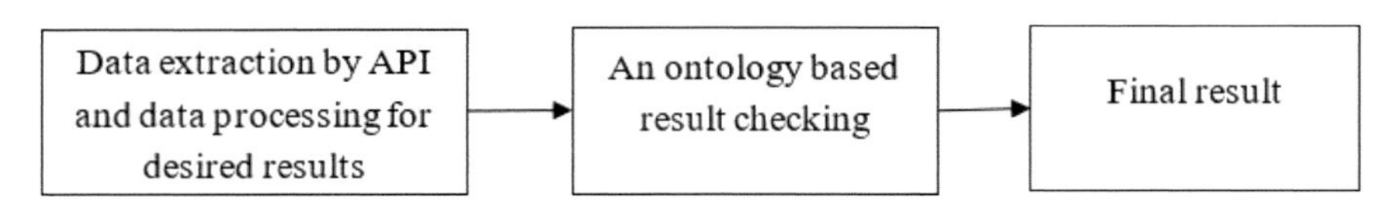

Figure 5.4 A type of hybrid data extraction.

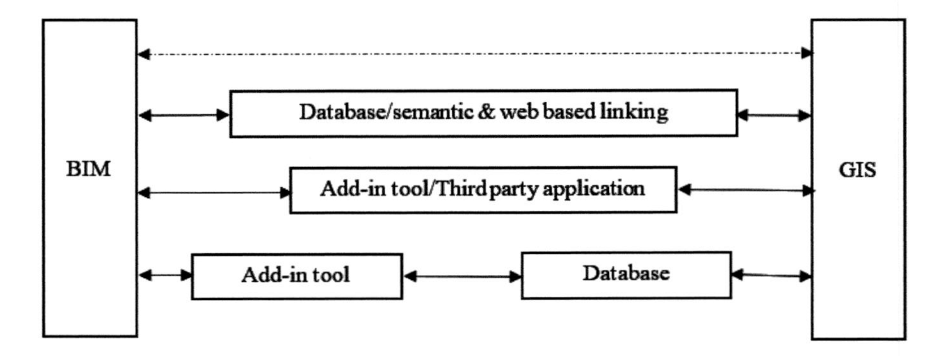

Figure 5.5 BIM and GIS integration for microgrid planning.

add-in tool development [20]. According to the customized data exchange requirement and capabilities of BIM and GIS applications, different ways of data exchange between BIM and GIS are possible, as partially detailed in Figure 5.5. Between BIM and GIS, required data can be extracted and exported. Semantic information system interlinking or data exchange may be either automatic, manual, semi-automatic, or a combination of these exchange processes [21, 22]. The smooth integration of heterogeneous data between BIM and GIS platforms is one of the major challenges for the developers of add-in tools [23].

5.5 Planning a Microgrid using BIM and GIS

An approach for microgrid planning using BIM and GIS is given in Figure 5.6. In this method, the microgrid planning tool is proposed in the BIM platform, and GIS is utilized to gather information. The qualitative comparison between the conventional 2D-CAD-based method and the proposed approach is given in Table 5.4.

5.6 Operation of Microgrid and Smart Built Environments

A promising methodology for smart grid management is the fusion of a BIM- and GIS-based microgrid planning model with sensor, weather, historic, and real-time data with an algorithm [24]. As shown in Figure 5.7, integrating smart grid components with related databases with BIM- and GIS-based microgrid model offers an automated platform for smart grid management [25].

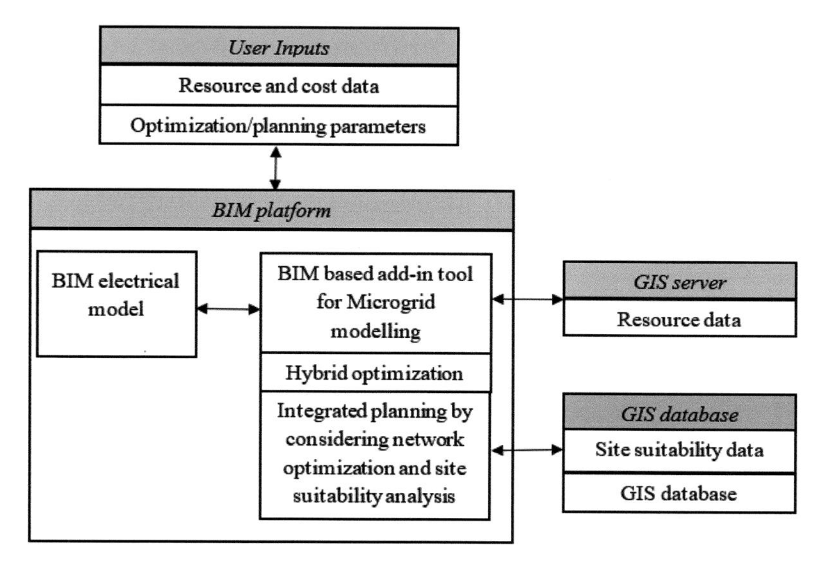

Figure 5.6 A practical approach for BIM- and GIS-based microgrid planning process.

Table 5.4 Qualitative comparison between the 2D-CAD method and the BIM- and GIS-based method microgrid planning.

Description of activity	Traditional 2D-CAD approach	BIM- and GIS-based method
Number of computational platforms	Multiple	Single
Occurrence of error	More	Less
Time spend for analysis	More	Less
GIS expertise	High	Less
Parametric modeling	No	Yes
Digitization of information	Less	More
Level of automation	Less	More
Sharing of data between platforms	Difficult	Easy
Interoperability issues	More	Less

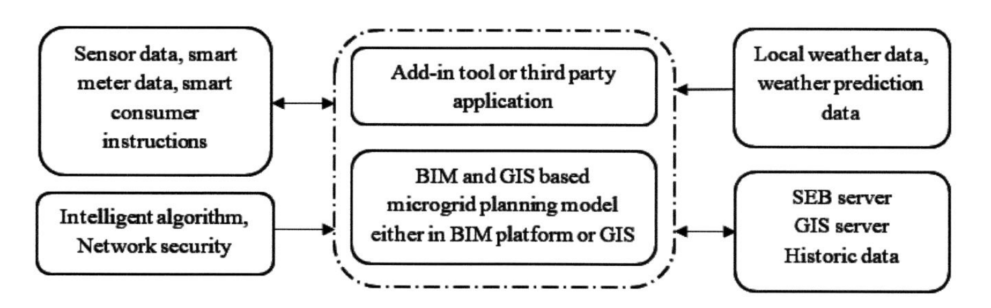

Figure 5.7 BIM and GIS integration for smart microgrid.

5.7 Major Applications of BIM-Based Microgrid Digitized Data at GIS Provisional-Level Planning

With BIM and GIS data integration, it is possible to conduct big data geo-analytics of multi-microgrid planning at the regional level. Single microgrid/ building energy consumption or statistical data is compared with both the average consumption in the district and the consumption level projected by the town planning policy. Feedback on further improvement of the microgrid/building design is given to individual low-energy building designers for necessary changes in their design according to regional planning goals [26]. Using BIM tools at microgrid/building level, energy conservation studies are possible, and the outcomes can be utilized to assess energy conservation measures at the city level [27]. The major applications of BIM for GIS-level electrical manipulations are shown in Figure 5.8, out of which management of electrical infrastructure and technological assessments are further discussed in the following sections. Using appropriate BIM-based tools, necessary data is to be exported to the GIS application incompatible formats such as SHP file formats with geo-coordinates and other required information. Combining GIS layers such as existing electrical network and power consumption data and BIM data required analysis can be conducted in the GIS platform. By

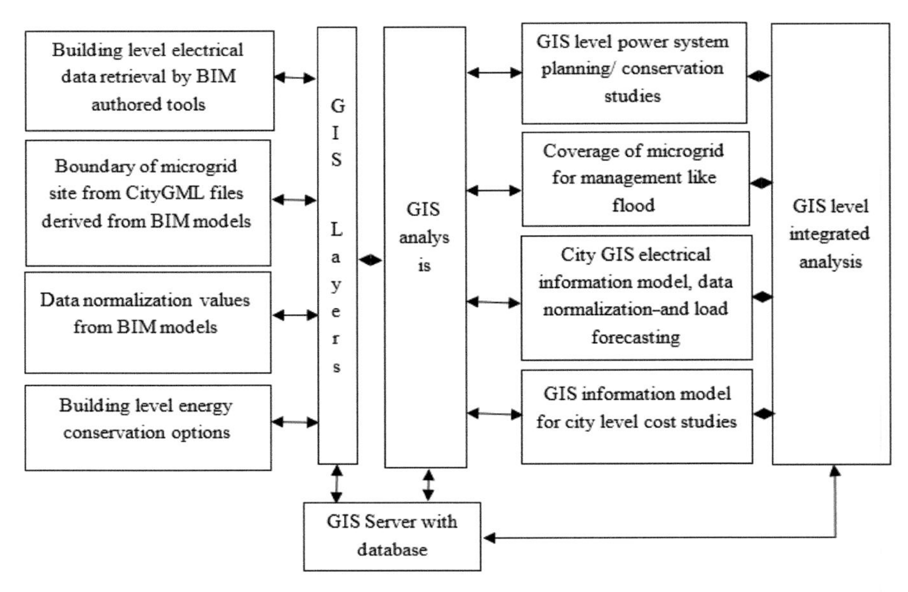

Figure 5.8 BIM and GIS combined applications for regional-level electrical system planning.

combining layers of different disciplines in GIS, the whole city integrated planning is possible [28].

Data normalization of the continuously varying load profile is possible by integrating BIM and GIS, which leads to a detailed, comprehensive national electrical energy consumption model. The same data standardization can be done for microgrid design BIM models, such as solar and wind installation costs, as shown in Figures 5.9 and 5.10.

Pre-construction BIM model is used to develop real-time power consumption models that closely locate where and how power consumption takes place in the building. Hence, different kinds of qualitative and quantitative load profile data can be transferred to GIS from both pre- and post-construction

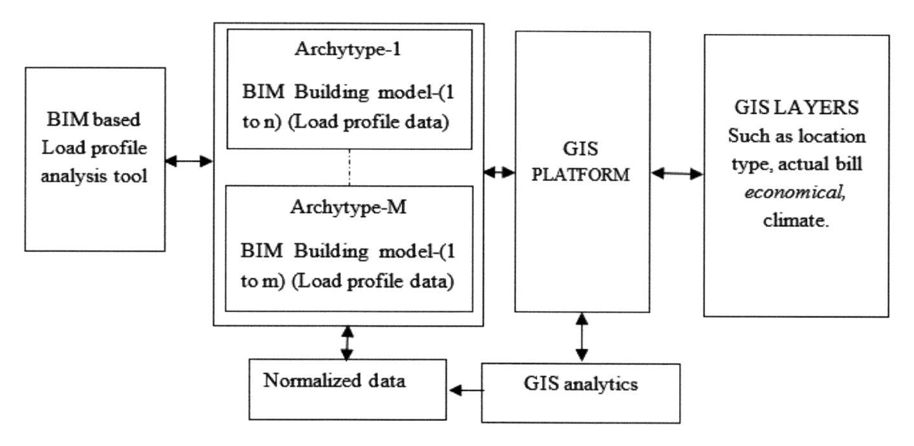

Figure 5.9 The process of load profile data standardization by BIM and GIS.

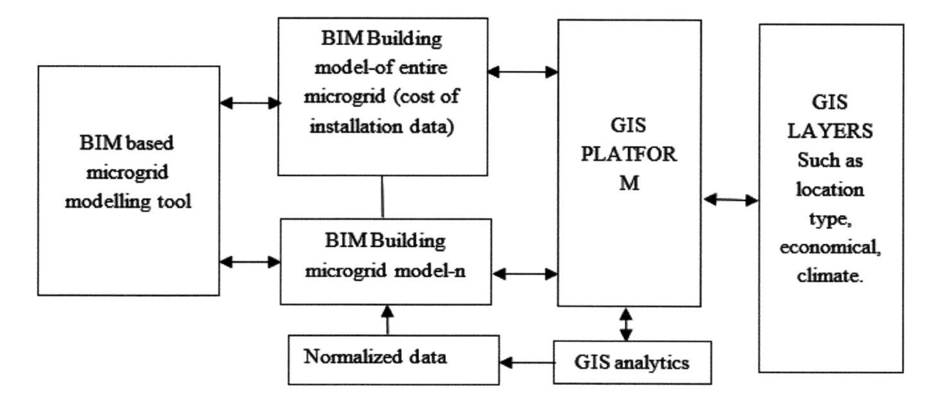

Figure 5.10 The process of microgrid-level standardization by BIM and GIS.

BIM models. By analyzing and normalizing each and every substantially important data, policies for energy conservation and smart built environment can be drawn by authorities and can generalize load pattern forecast for each archetype at the provincial level. National and average regional imprecisions can be narrowed down by combining monthly billing, sub-metered data details and BIM data (detailed housing information), and geographical and other related details. This method minimizes exhaustive surveys and reduces the time for analysis.

5.8 Management

BIM- and GIS-based integration can accomplish proactive planning for electrical supply management during disasters. BIM and GIS combined district energy information modeling can provide coverage of every electrical generation source, transmission and distribution network details, and their connection details. Following the method in Figure 5.11, an electricity management procedure can be developed using GIS applications during a flood situation. The disaster management team can know that the area has an emergency supply from the microgrid. From BIM geographical location, project boundary

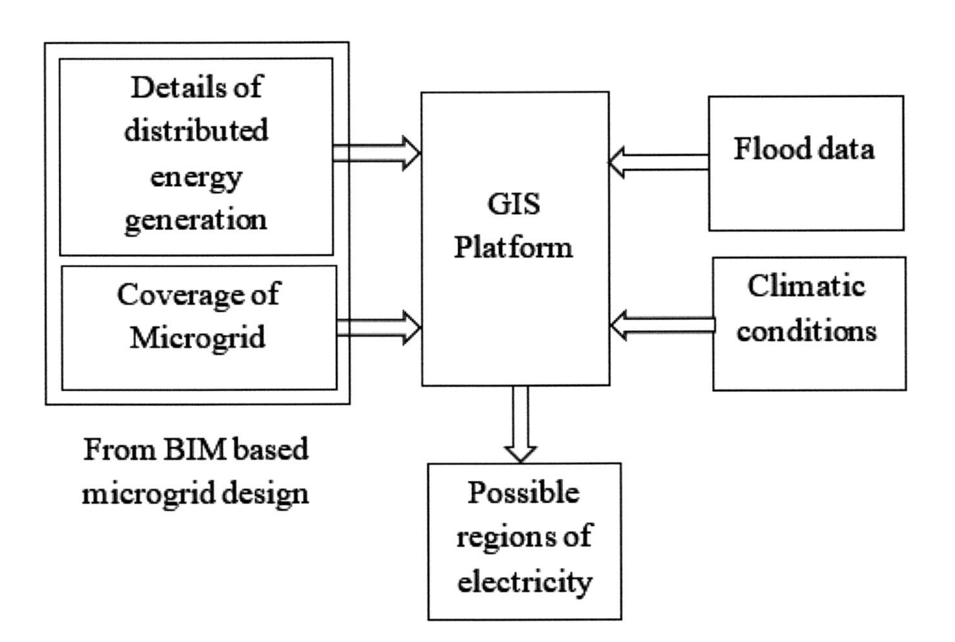

Figure 5.11 Flood time microgrid data use for management.

cost details, and all the required information related to microgrid is exported to GIS by add-in tools/BIM file exporting options. By combining microgrid data with flood layers, proactive planning and operational schedules at different flood levels can be prepared. The same can be deployed during a flood for energy supply. Thus, it is possible to add vital information, without any errors and without the need of any specialist, to every construction project leading to a detailed city information model for energy management.

5.9 Technology Assessment

The town planning department is required to assess construction projects for energy efficiency and give feedback to designers according to the specified conservation measures. Processing a massive volume of construction documents for related information in the conventional 2D format is laborious. According to country planning requirements, BIM and GIS technology can be effectively used to assess alternative technologies for building systems. Suppose the data related to energy conservation possibilities exists. In that case, the town planners can search for priority buildings that can offer a positive return on investments throughout the lifespan of the building. Figure 5.12 shows the block diagram for technological assessment for energy usage comparison using BIM and GIS. A city energy planner can assess the effectiveness of the energy conservation measures in a building using customized tools, which can then be compared with regional-level GIS data. Accordingly, he can give feedback to the owner of the building related to energy utilization.

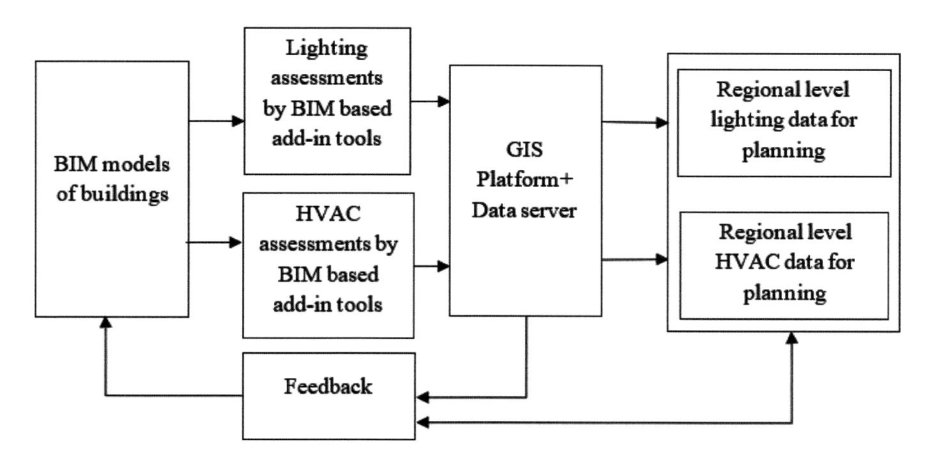

Figure 5.12 Technology assessment by BIM and GIS integration.

5.10 Requirements for Implementing BIM- and GIS-Based Microgrid Planning

The major requirements for the implementation of integrated BIM and GIS platform for the planning of microgrid are as follows:

i. Government regulations and modeling standardization
ii. Web-platform-based microgrid data exchange and continuous evaluations
iii. Acceptance and training by firms
iv. Freeware applications
v. Regional-level strategies for integrating BIM and GIS data and mitigating errors in data exchange

5.11 Summary

Recent progress in information and communications technologies (ICT) has empowered us to collect, integrate, process, and manage the infrastructural information systematically and locally, which was difficult in the past. Unlike the conventional microgrid planning and design approach, the integrated BIM-GIS approach retrieves related data in a consolidated platform for design, modeling, and analysis. This chapter has provided a summary of the potential role of BIM and the GIS integrated approach for microgrid planning and its applications. Integrating the semantic information system of BIM with GIS can be advantageous for microgrid optimal planning and management by considering component-level and network-level impact on solar plants. BIM provides building-level electrical data and optimization studies, and GIS integrates the building-level data for city-level studies. This integrated microgrid modeling approach, cost-effective design, ease of information capture for analysis, smart management, better visualization, and communication can be achieved. The major task of this proposed approach is to gather and integrate heterogeneous data and develop an automated tool according to a large number of manual codes and regional regulations. Using BIM and GIS for microgrid planning by combining efficient optimization algorithms, artificial intelligence, and big data analytics is a promising research field.

References

[1] A. Hanna, F. Boodai and M. El Asmar, "State of Practice of Building Information Modeling in Mechanical and Electrical Construction

Industries", Journal of Construction Engineering and Management, vol. 139, no. 10, 2013, 04013009. Available: 10.1061/(asce)co.1943-7862.0000747 [Accessed 29 June 2021].

[2] J. Farooq, P. Sharma and S. R. Kumar, "Applications of Building Information Modeling in Electrical Systems Design", Journal of Engineering Science and Technology Review, vol. 10, no. 6, pp. 119–128, 2017. Available: 10.25103/jestr.106.16 [Accessed 29 June 2021].

[3] "Electric Utilities Design & Engineering | Utility Network Management", Esri.com, 2021. [Online]. Available: https://www.esri.com/en-us/industries/electric/business-areas/design-engineering. [Accessed 30 June 2021].

[4] Y. Song *et al.*, "Trends and Opportunities of BIM-GIS Integration in the Architecture, Engineering and Construction Industry: A Review from a Spatio-Temporal Statistical Perspective", ISPRS International Journal of Geo-Information, vol. 6, no. 12, p. 397, 2017, pp. 1–32. Available: 10.3390/ijgi6120397.

[5] I. Baht, P. Nicolae and M. Nicolae, "Impact of Weather Forecasts and Green Building on Micro Grid Energy Management System", 2019 International Conference on Electromechanical and Energy Systems (SIELMEN), 2019, pp. 1–6. Available: 10.1109/sielmen.2019.8905846 [Accessed 30 June 2021].

[6] J. Farooq, R. Pachauri, R. Sreerama Kumar and P. Sharma, "An Add-in Tool for BIM-Based Electrical Load Forecast for Multi-Building Microgrid Design", Proceedings of International Conference on Artificial Intelligence, Smart Grid and Smart City Applications, pp. 57-67, 2020. Available: 10.1007/978-3-030-24051-6_6 [Accessed 30 June 2021].

[7] R. Jang and W. Collinge, "Improving BIM Asset and Facilities Management Processes: A Mechanical and Electrical (M&E) Contractor Perspective", Journal of Building Engineering, vol. 32, 2020, 101540. Available: 10.1016/j.jobe.2020.101540 [Accessed 30 June 2021].

[8] J. Wang, X. Wang, W. Shou, H. Chong and J. Guo, "Building Information Modeling-Based Integration of MEP Layout Designs and Constructability", Automation in Construction, vol. 61, pp. 134–146, 2016. Available: 10.1016/j.autcon.2015.10.003 [Accessed 30 June 2021].

[9] Monika, D. Srinivasan and T. Reindl, "GIS as a Tool for Enhancing the Optimization of Demand Side Management in Residential Microgrid", 2015 IEEE Innovative Smart Grid Technologies - Asia (ISGT ASIA), 2015, pp. 1–6. Available: 10.1109/isgt-asia.2015.7387041 [Accessed 30 June 2021].

[10] M. Wang, Y. Deng, J. Won and J. Cheng, "An Integrated Underground Utility Management and Decision Support Based on BIM and GIS",

Automation in Construction, vol. 107, 2019, 102931. Available: 10.1016/j.autcon.2019.102931 [Accessed 3 July 2021].

[11] E. Karan and J. Irizarry, "Extending BIM Interoperability to Pre-Construction Operations using Geospatial Analyses and Semantic Web Services", Automation in Construction, vol. 53, pp. 1–12, 2015. Available: 10.1016/j.autcon.2015.02.012 [Accessed 4 July 2021].

[12] A. Wahab and J. Wang, "Factors-Driven Comparison Between BIM-Based and Traditional 2D Quantity Takeoff in Construction Cost Estimation", Engineering, Construction and Architectural Management, Vol. 29, Issue No. 2, pp. 702–715. Available: 10.1108/ecam-10-2020-0823 [Accessed 5 July 2021].

[13] B. Wang, C. Yin, H. Luo, J. Cheng and Q. Wang, "Fully Automated Generation of Parametric BIM for MEP Scenes Based on Terrestrial Laser Scanning Data", Automation in Construction, vol. 125, 2021, 103615. Available: 10.1016/j.autcon.2021.103615 [Accessed 5 July 2021].

[14] A. Ismail, K. Ali and N. Iahad, "A Review on BIM-Based Automated Code Compliance Checking System", 2017 International Conference on Research and Innovation in Information Systems (ICRIIS), 2017, pp. 1–6. Available: 10.1109/icriis.2017.8002486 [Accessed 5 July 2021].

[15] R. Nizam, C. Zhang and L. Tian, "A BIM Based Tool for Assessing Embodied Energy For Buildings", Energy and Buildings, vol. 170, pp. 1–14, 2018. Available: 10.1016/j.enbuild.2018.03.067 [Accessed 30 June 2021].

[16] H. Liu, M. Lu and M. Al-Hussein, "Ontology-Based Semantic Approach for Construction-Oriented Quantity Take-Off from BIM Models in the Light-Frame Building Industry", Advanced Engineering Informatics, vol. 30, no. 2, pp. 190-207, 2016. Available: 10.1016/j.aei.2016.03.001 [Accessed 30 June 2021].

[17] "Revit Platform Technologies | Autodesk Developer Network", Autodesk.com, 2021. [Online]. Available: https://www.autodesk.com/developer-network/platform-technologies/revit. [Accessed 07 July 2021].

[18] S. Lee, K. Kim and J. Yu, "BIM and Ontology-Based Approach for Building Cost Estimation", Automation in Construction, vol. 41, pp. 96–105, 2014. Available: 10.1016/j.autcon.2013.10.020 [Accessed 7 July 2021].

[19] S. Su, Q. Wang, L. Han, J. Hong and Z. Liu, "BIM-DLCA: An Integrated Dynamic Environmental Impact Assessment Model for Buildings", Building and Environment, vol. 183, 2020, 107218. Available: 10.1016/j.buildenv.2020.107218 [Accessed 7 July 2021].

[20] J. Zhu, P. Wu, M. Chen, M. Kim, X. Wang and T. Fang, "Automatically Processing IFC Clipping Representation for BIM and GIS Integration

at the Process Level", Applied Sciences, vol. 10, no. 6, pp. 1–19, 2009, 2020. Available: 10.3390/app10062009.

[21] H. Wang, Y. Pan and X. Luo, "Integration of BIM and GIS in Sustainable Built Environment: A Review and Bibliometric Analysis", Automation in Construction, vol. 103, pp. 41–52, 2019. Available: 10.1016/j.autcon.2019.03.005 [Accessed 30 June 2021].

[22] Z. Ma and Y. Ren, "Integrated Application of BIM and GIS: An Overview", Procedia Engineering, vol. 196, pp. 1072–1079, 2017. Available: 10.1016/j.proeng.2017.08.064 [Accessed 30 June 2021].

[23] Y. Bai, P. Zadeh, S. Staub-French and R. Pottinger, "Integrating GIS and BIM for Community-Scale Energy Modeling", International Conference on Sustainable Infrastructure 2017, pp. 185–196. Available: 10.1061/9780784481196.017 [Accessed 3 July 2021].

[24] M. Amini, H. Arasteh and P. Siano, "Sustainable Smart Cities Through the Lens of Complex Interdependent Infrastructures: Panorama and State-of-the-Art", Studies in Systems, Decision and Control, pp. 45-68, 2018. Available: 10.1007/978-3-319-98923-5_3 [Accessed 3 July 2021].

[25] J. Woo and C. Menassa, "Virtual Retrofit Model for Aging Commercial Buildings in a Smart Grid Environment", Energy and Buildings, vol. 80, pp. 424–435, 2014. Available: 10.1016/j.enbuild.2014.05.004 [Accessed 7 July 2021].

[26] W. Wu, X. Yang and Q. Fan, "GIS-BIM Based Virtual Facility Energy Assessment (VFEA)—Framework Development and Use Case of California State University, Fresno", Computing in Civil and Building Engineering (2014), 2014. Available: 10.1061/9780784413616.043 [Accessed 7 July 2021].

[27] A. Schlueter, P. Geyer and S. Cisar, "Analysis of Georeferenced Building Data for the Identification and Evaluation of Thermal Microgrids", Proceedings of the IEEE, vol. 104, no. 4, pp. 713-725, 2016. Available: 10.1109/jproc.2016.2526118 [Accessed 7 July 2021].

[28] M. Marzouk and A. Othman, "Planning Utility Infrastructure Requirements for Smart Cities using the Integration Between BIM and GIS", Sustainable Cities and Society, vol. 57, 2020, 102120. Available: 10.1016/j.scs.2020.102120 [Accessed 7 July 2021].

Chapter 6

Comparative Study on the Thermo-Hydraulic Performance of Corrugated and Impinging Jet Solar Air Heater

Siddhita Yadav[1*] and R. P. Saini[2]

[1]Research Scholar, HRED, IIT Roorkee, India
[2]Professor, HRED, IIT Roorkee, India
*Corresponding author: syadav@ah.iitr.ac.in
https://orcid.org/ 0000-0001-5324-3768

Abstract

Energy demand is increasing drastically with the increasing population. Thus, fossil fuels are depleting faster, and these are non-replenishable resources that may result in energy scarcity worldwide. Fossil fuels can be replaced with renewable energy sources to meet environmentally friendly energy demands. Solar energy seems to be the most feasible option among all alternative energy sources. Solar energy can be used for heating applications using solar air heaters (SAHs). However, there is still a lot to improve. In this study, the thermal behavior of a solar air heater with a corrugated plate (CSAH) and with jet impingement technique (SAHJI) has been analyzed and compared with a smooth plate solar air heater (SSAH) under identical conditions of operation. ANSYS FLUENT tool is adapted for present simulations. The 3D model of CSAH and SAHJI is generated with similar duct sizes. The RNG *k*-turbulence model has been used to calculate the transport equations. In order to carry out a comparative study for the performance of CSAH, SAHJI, and SSAH, heat transfer in terms of Nusselt number (Nu) and pumping power in terms of friction factor (ff) have been considered as 1000 W/m^2 for Reynolds number varying from 3700 to 16,500. For CSAH and SAHJI, the results show a considerable increase in heat transfer of

1.47 and 1.65 times, respectively, when compared to SSAH. The maximum thermo-hydraulic performance parameter is obtained as 1.18 for SAHJI. For the range of parameters considered under the present study, the thermo-hydraulic performance of SAHJI is found to be better than CASH and SSAH under similar conditions.

6.1 Introduction

The major problem for humanity's evolution is energy production and preservation. Solar energy is the best form of renewable energy in terms of cost and environmental impact, and it is a readily available, clean energy source on the planet [1, 2]. According to applications, the energy coming from the sun can be converted into electrical energy using PV cells or thermal energy using an appropriate fluid. One of the most efficient systems for utilizing solar thermal energy is a solar air heater [3]. SAH contributes to the literature by meeting the residential heating and cooling and crop drying requirements. Despite its numerous benefits over liquid heating systems (no leaks, freezing, or rusting), SAH has the disadvantage of the low heat-absorbing ability of air from a hot absorber surface. Performance of SAH can be improved through its modification like artificial roughness, reduction in heat losses to the surrounding, and increasing duct turbulence [4–6].

Using the jet impingement approach, researchers have conducted numerous experiments on SAH, but due to software restrictions and complex fluid flow, CFD simulation is used only in certain investigations. The jet impingement technique was first introduced by Kercher and Tabakoff [7] to investigate the heat transfer characteristics of circular inline jets, and later this technique was used for SAH. SAHJI has advantages of high heat transfer rate, low pumping power, low-pressure losses, and cost-effectiveness. Further, Metzger *et al.* [8] studied the thermal characteristics of impinging jets and observed that inline jets outperform the staggered jet design. Choudhury and Garg [9] explored the mathematical energy equations for SAHJI experimentally. The thermal and hydraulic characteristics of SAHJI were reported by Chauhan and Thakur [10], who took into account the influence of distinct jet patterns. Later, Chauhan *et al.* [11] improved different variables of the jet-plate-like, jet diameter, and spanwise pitch using the PSI method. Belusko *et al.* [12] discussed the performance analysis and application of an unglazed SAHJI, and Zukowski [13] presented the effect of spanwise pitch and slots [14]. Yadav and Saini [15–17] have numerically analyzed the thermo-hydraulic performance of SAHJI with different jet geometries and jet parameters like jet diameter ratio and jet height ratio.

Using the corrugated plate as an absorber is also beneficial for increasing the turbulence inside the air duct and the absorber area to trap more heat from the sun [18]. Gao *et al.* [19] analyzed thermal performance of cross-corrugated solar air heater analytically and experimentally. This study concluded that the thermal efficiency of CSAH is 10% higher than SSAH for the considered parametric range. El-Sebaii *et al.* [20] compared the thermal performance of V-CSAH and double pass SSAH numerically and found out that V-SAH is more efficient than double pass SSAH. Varol and Oztop [21] conducted numerical investigations on the natural convection heat transfer characteristics in the air gap between the wavy absorber surface and glass cover of a solar air heater. The minimum value of the Nusselt number was obtained for the configuration having an aspect ratio of 4, and the normalized wavelength of 4. Wandong *et al.* [6] used a corrugated absorber plate of a thickness is 0.15 mm for room heating. The porosity was about 10%. The perforation diameter was about 4 mm. The corrugation angle of the absorber plate was 45°. Karim and Hawlader [22] experimentally investigated the thermal performance characteristics of solar air heaters fitted with a V-corrugated absorber plate used for drying applications. The analysis shows that the V-corrugations provide 12% higher efficiency than a flat plate absorber. Gao *et al.* [23] carried out an experimental and analytical study to determine the heat transfer effectiveness of cross-corrugated solar air heater. For Type I collector, the corrugations on the bottom plate was kept across the flow while the corrugations on the absorber plate was along the flow direction. For Type II collector, the arrangements of bottom and absorber plates were reversed. The maximum thermal efficiency for Type I, Type II, and flat plate collector systems were found to be about 58.9%, 60.3%, and 48.6%, respectively.

According to a set of studies, previous research has primarily concentrated on experimental studies, with only a few investigators using the CFD tool to analyze solar air heaters. CFD can be used to investigate SAH performance to reduce the cost and time in the investigations [24, 25]. Keeping this in mind, the current study uses ANSYS-FLUENT-18.1 to conduct a CFD-based investigation of the thermal behavior of the CSAH and SAHJI. CFD results are validated by numerical equations and experimental results available in the literature. Using present analysis results, a comparative investigation has been attempted on the performance of CSAH and SAHJI with SSAH using the same operating conditions. The comparison of heat transfer behavior of CSAH and SAHJI has been made with SSAH. The current investigation findings can be helpful for further investigations on solar air heaters.

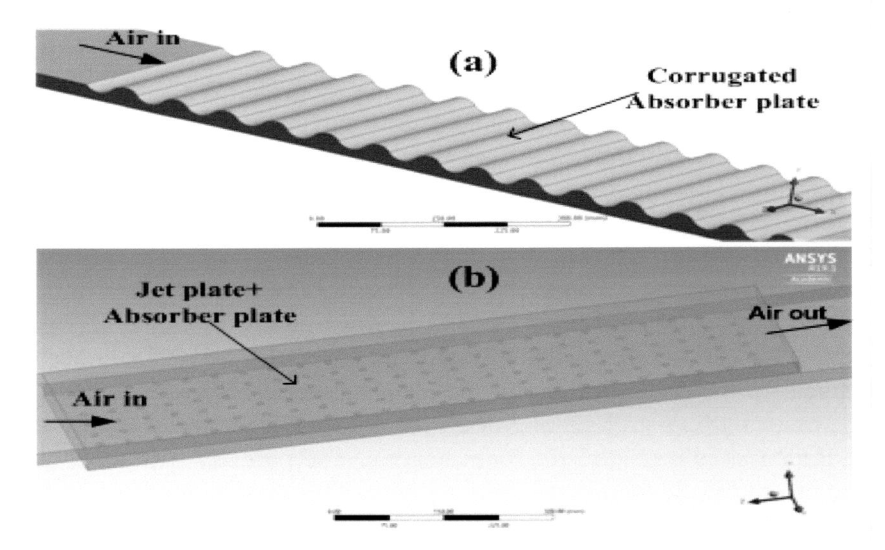

Figure 6.1 3D modeling of (a) corrugated (b) impinging jet solar air heater.

6.2 Computational Methodology

6.2.1 Geometry and meshing

This study is conducted on a SAH duct with 0.3 m (width) × 0.025 m (depth) and 1 m in length with an entry length of 575 mm and exit length of 400 mm as per ASHRAE guidelines [26, 27]. The duct has an absorber plate, and jet plate made up of aluminum. Sidewalls and backplates are insulated. Based on previous investigations, geometrical parameters for the jet plate are taken as streamwise, and spanwise pitch ratios are 1.67 and 0.867, respectively, for a circular jet diameter ratio of 0.065 [11]. The characteristic-height ratio of 3.5 and non-dimensional wavelength of 2 are adopted as the geometrical parameter of a corrugated absorber plate in CSAH [28], as shown in Figure 6.1. ANSYS Workbench has been used to generate a 3D model of SAHJI and CSAH, and ANSYS-ICEM is used to generate the mesh. Present 3D geometry is discretized with a fine, unstructured, and non-uniform grid for the meshing.

GIT has been used to identify the optimum number of elements in order to reduce time, achieve grid convergence, and improve numerical stability. For GIT, the number of elements ranged from 1.4 million to 1.9 million. The meshing of SAHJI with 1.75 million elements is proven to be the best suited for this design so that an error in results (Nu) is less than 1% above this value. To analyze the flow field at the inlet and outlet, inflation has been provided

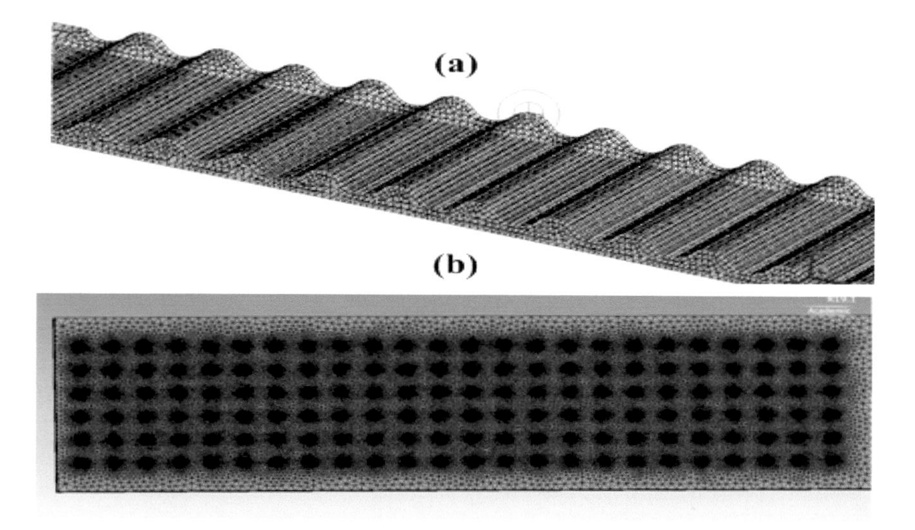

Figure 6.2 (a) CSAH and (b) SAHJI mesh.

with a growth rate of 1.02. Figure 6.2(a) and (b) show mesh for CSAH and SAHJI, respectively.

6.2.2 Boundary conditions

The existing study's 3D model consists of a rectangle-shaped duct with an entrance region where air enters the duct, a test section examining heat transfer characteristics, and an exit region where hot air exits into the atmosphere. The test section's upper wall is a hot absorber surface with a 1000 W/m^2 evenly distributed heat flux. Computations were performed for Reynolds numbers (Re) ranging from 3700 to 16,500 (six steps). Inside the duct, the working fluid was referred to as ambient air. Aluminum was chosen as the material for the jet and absorber plate.

Required boundary conditions were provided based on real operations for the 3D computational model using FLUENT solver for simulation. Boundary conditions to calculate the governing equations are mentioned in Table 6.1. In order to calculate the continuity, momentum, and energy equations, the RNG k-ε turbulence model is preferred to simulate the flow inside the SAH duct [15, 29, 30]. With respect to other turbulence models, the two equations k-epsilon turbulence approach integrates statistical correctness and saves calculation time. The results obtained with the smooth plate solar air heater experiment agree with CFD results [14, 15].

Table 6.1 Applied boundary conditions on flow domain.

Location	Boundary conditions	Type
Air entry	Velocity inlet	Dirichlet
Air exit	Pressure outlet	Dirichlet
Absorber plate	Wall	No-slip
Sidewalls	Wall	No-slip, adiabatic
Contact surfaces	Interface	Symmetry

6.3 Results and Discussions

The outcomes using the equations and experimental results provided in the literature have been used to validate the CFD model of SSAH [30], assuming similar operational circumstances. The thermal characteristics of SAHJI and CSAH are compared with those of plain SAH, and the outcomes of this comparison are presented in the following section. Contours show the variance of temperature and pressure with rising Re values.

6.3.1 Validation

Between results obtained from CFD analysis and the Dittus–Boelter equation [25], the average absolute deviation of Nu is 4.68%. CFD and experimental results could be attributed to simplifying assumptions of higher-order equations and rounds of errors used to run the simulation. Another reason for deviation may be the assumed conditions and actual experimental conditions. The results from earlier experimental data show close agreement with the findings from CFD data (Figure 6.3).

6.3.2 Temperature and pressure distribution

Figures 6.4 and 6.5 show the temperature and pressure distribution for SSAH, CSAH, and SAHJI duct, respectively, for various Re values from 3700 to 16,500. It can be said that whenever the Re value rises, the temperature of the absorber surface decreases while the pressure within the duct rises. The temperature of the absorber and the pressure within the duct rise as a result of the formation of a laminar sub-layer on the absorber surface.

The corrugated profile has a bigger surface area for heat exchange and enhances turbulence within the duct for CSAH. SAHJI, on the other hand, has a higher temperature at the plate's edges and a lower temperature where the jet impacts the heated plate. The impact of the jet on the absorber surface disturbs the laminar sub-layer, and heat exchange is increased between the absorber and air inside the duct. As a result, the heat exchange rate for SAHJI and CSAH is more than that of SSAH.

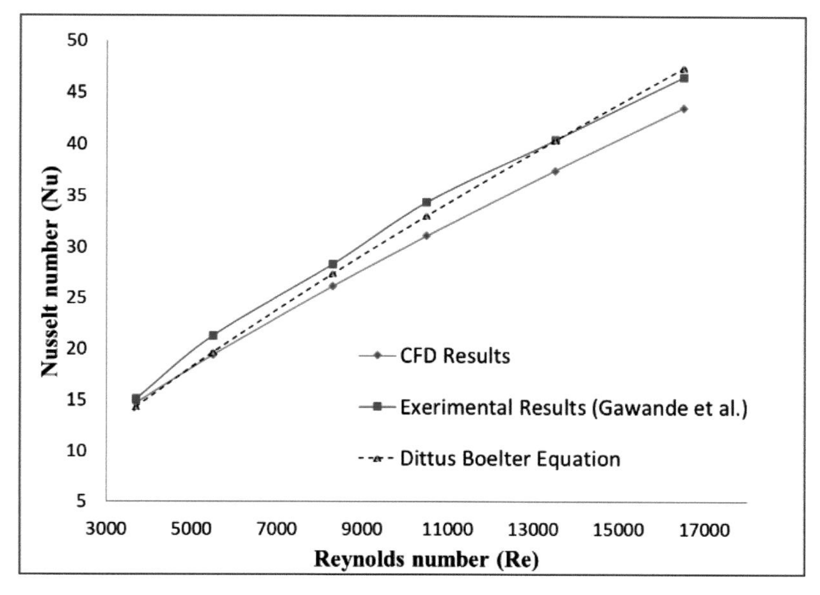

Figure 6.3 Validation of CFD results.

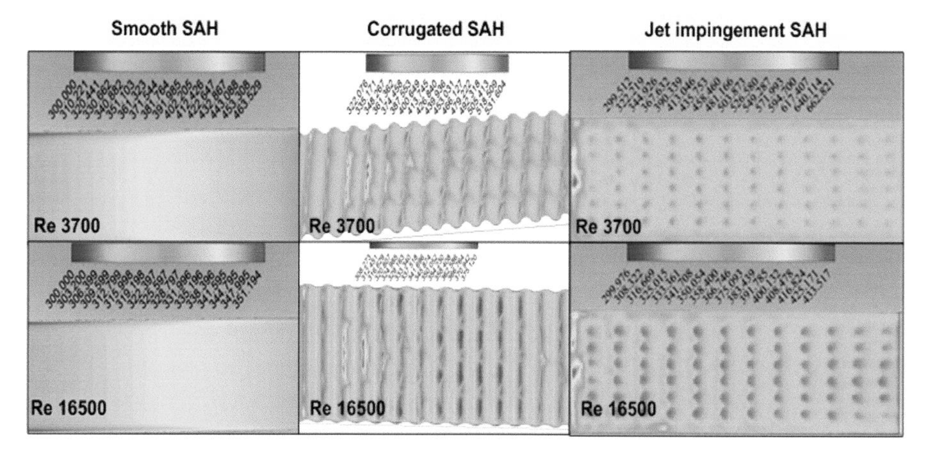

Figure 6.4 Temperature (K) distribution for SSAH, CSAH, and SAHJI at Re = 3700 and 16,500.

6.3.3 Flow pattern inside the duct

Figures 6.6(a) and (b) illustrate CSAH and SAHJI streamlines within the duct, respectively. For CSAH, the air follows a corrugated path. It absorbs heat from the surface, while in the case of SAHJI, air absorbs heat by cross-flow through a jet plate inside the duct. This impact of the jet on the absorber

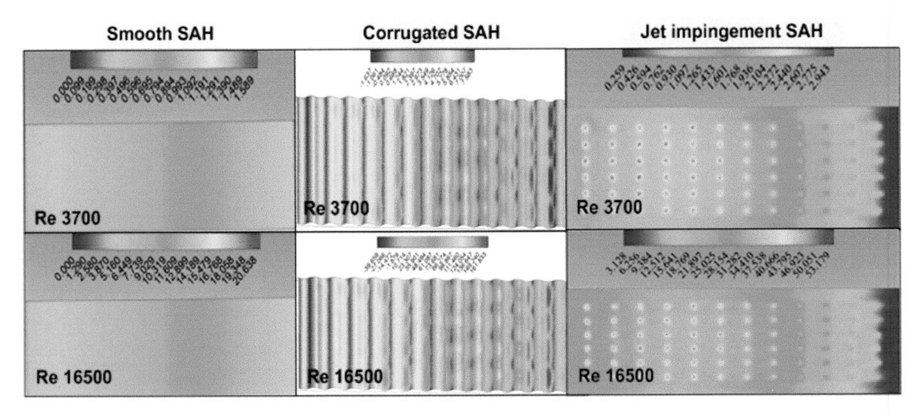

Figure 6.5 Pressure distribution (Pa) for SSAH, CSAH, and SAHJI at Re = 3700 and 16,500.

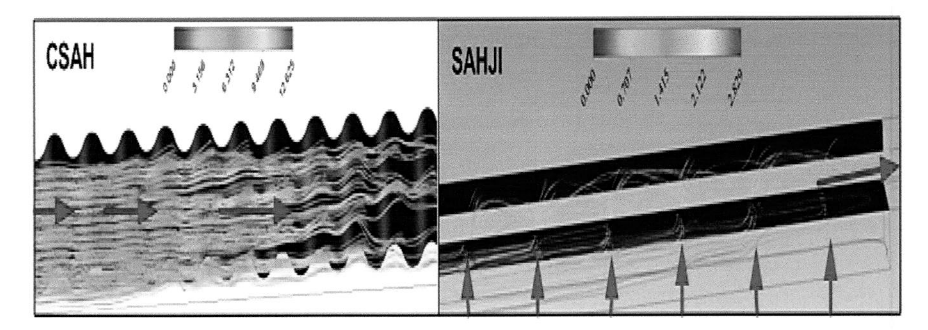

Figure 6.6 Contours of streamlines inside the duct for (a) CSAH and (b) SAHJI.

plate increases the heat energy transferred from the air to the absorber plate by disturbing the laminar sub-layer and causing turbulence within the duct. The time it takes to transfer the heat is minimized when the incoming air velocity rises. As a result, heat transfer is reduced when the inlet air velocity is very high. SAHJI and CSAH absorb more heat than SAH because the turbulence within the duct is increased.

6.3.4 Effect of re on heat transfer and friction factor

Figure 6.7(a) shows that Nu increases while ff drops as Re grows. For the specified range of Re, SAHJI has the largest Nu range, while CSAH has the largest ff range. As the amount of heat transferred increases, the amount of pumping power is required. Figure 6.7(b) illustrates that at Re 16,500, the maximum enhancements in heat transfer (NNE) for CSAH and SAHJI are

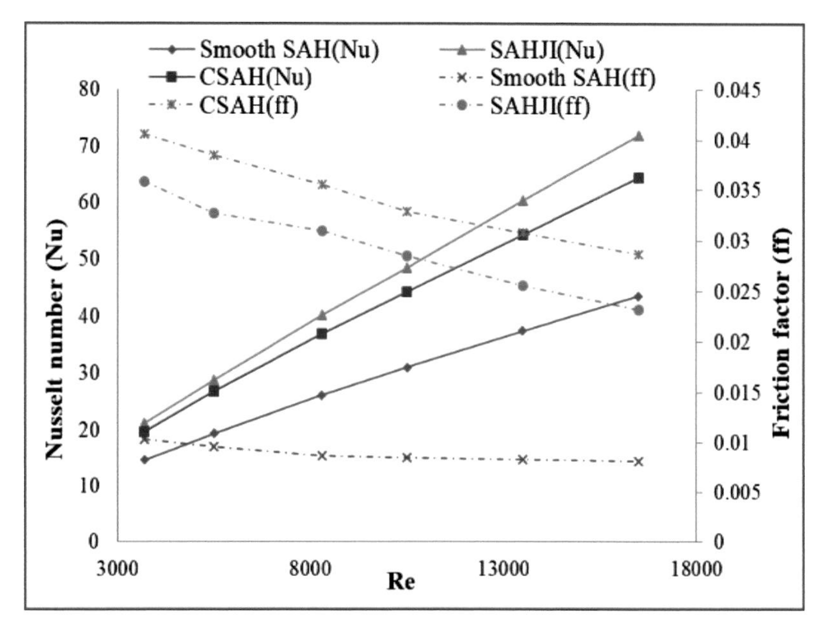

Figure 6.7(a) Re variation with Nu and ff.

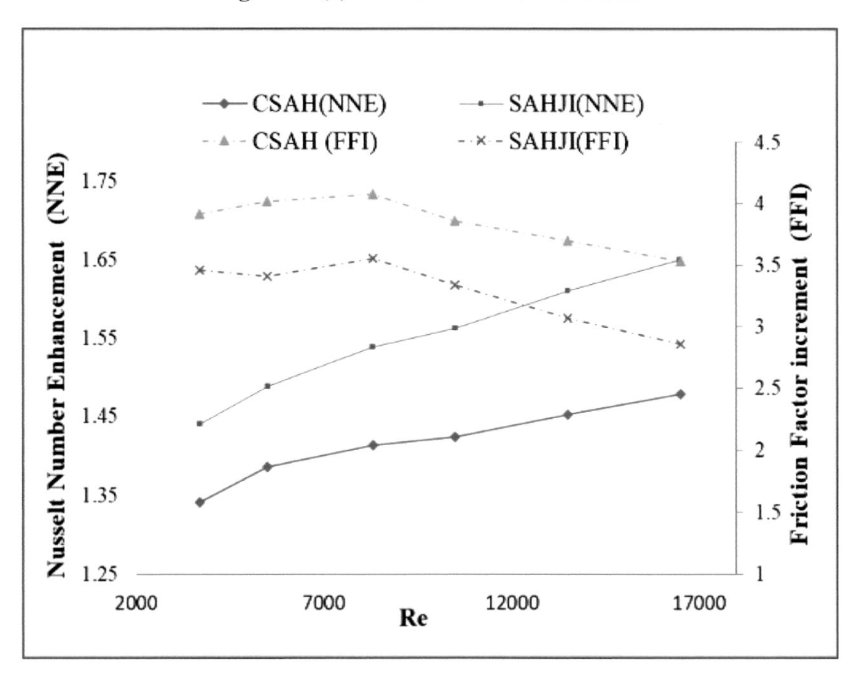

Figure 6.7(b) Re variation with NNE and friction FFI.

1.65 and 1.47, respectively, and the corresponding friction factor increment (FFI) is 3.4, which is 4.47 times that SSAH. For SAHJI, the upper limit of the thermo-hydraulic performance parameter [THPP = (NNE)/(FFI)1/3] was found to be 1.18. The value of THPP is more than 1.0, indicating that heat transfer dominates the friction factor for a given set of parameters. Thus, heat transfer characteristics of SAHJI are better than CSAH and SSAH.

6.4 Conclusion

Using ANSYS18.1, a numerical CFD simulation was conducted to investigate the thermo-hydraulic behavior of a CSAH, SAHJI, and SSAH. CFD findings demonstrate good agreement with reported earlier mathematical models and experimental results for the solar air heater under consideration in this study. According to the evaluation, the CFD tool is a reliable technique for studying SAH.

In the range of parameters investigated, the simulation results show that Nusselt number enhancement increases with Reynolds number from 1.44 to 1.65 for SAHJI and from 1.33 to 1.47 times for CSAH. The thermal exchange order is SSAH < CSAH < SAHJI in ascending order. As a result, in terms of heat transfer, SAHJI is determined to be more efficient than CSAH and SSAH.

As the Reynolds number rises, the friction factor drops. For SAHJI and CSAH, the friction factor increased from 3.46 to 2.85 and 3.92 to 3.53, respectively. The friction factor determines the amount of power necessary to move the air through the duct, which is written as SSAH < SAHJI < CSAH.

Further, the maximal THPP value is obtained as 1.18 for SAHJI corresponds to Reynolds number 16,500 within the investigated set of parameters. Because the THPP is higher than 1.0, heat exchange takes precedence over friction inside the duct. With lower pumping power, the overall heat transfer performance of impinging jet solar air heaters (SAHJI) is superior to CSAH and SSAH.

References

[1] A. Gautam and R. P. Saini, "Thermal and hydraulic characteristics of packed bed solar energy storage system having spheres as packing element with pores," J. Energy Storage, vol. 30, March 2020, Art. no. 101414, doi: 10.1016/j.est.2020.101414.

[2] G. Saini, A. Kumar, R. P. Saini, and G. Dwivedi, "Effect of number of stages on the performance characteristics of modified Savonius hydrokinetic turbine," Ocean Eng., vol. 217, October 2020, Art. no. 108090, doi: 10.1016/j.oceaneng.2020.108090.

[3] R. Nadda, A. Kumar, and R. Maithani, "Efficiency improvement of solar photovoltaic/solar air collectors by using impingement jets: A review," Renew. Sustain. Energy Rev., vol. 93, no. 2017, pp. 331–353, 2018, doi: 10.1016/j.rser.2018.05.025.
[4] A. D. Gupta and L. Varshney, "Performance prediction for solar air heater having rectangular sectioned tapered rib roughness using CFD," Therm. Sci. Eng. Prog., vol. 4, pp. 122–132, 2017, doi: 10.1016/j.tsep.2017.09.005.
[5] V. B. Gawande, A. S. Dhoble, D. B. Zodpe, and S. Chamoli, "Analytical approach for evaluation of thermo hydraulic performance of roughened solar air heater," Case Stud. Therm. Eng., vol. 8, pp. 19–31, 2016, doi: 10.1016/j.csite.2016.03.003.
[6] W. Zheng, H. Zhang, S. You, Y. Fu, and X. Zheng, "Thermal performance analysis of a metal corrugated packing solar air collector in cold regions," Appl. Energy, vol. 203, pp. 938–947, 2017, doi: 10.1016/j.apenergy.2017.06.016.
[7] D. M. Kercher and W. Tabakoff, "Heat transfer by a square array of Round air jets impinging perpendicular to a flat surface including the effect of spent air," ASME-Paper 69-GT-4, 1969.
[8] D. E. Metzger, L. W. Florschuetz, D. I. Takeuchi, R. D. Behee, and R. A. Berry, "Heat transfer characteristics for inline and staggered arrays of circular jets with cross-flow of spent air," J. Heat Transfer, vol. 101, no. 3, pp. 526–531, 1979, doi: 10.1115/1.3451022.
[9] C. Choudhury and H. P. Garg, "Evaluation of a jet plate solar air heater," Sol. Energy, vol. 46, no. 4, pp. 199–209, 1991, doi: 10.1016/0038-092X(91)90064-4.
[10] R. Chauhan and N. S. Thakur, "Investigation of the thermohydraulic performance of impinging jet solar air heater," Energy, vol. 68, pp. 255–261, 2014, doi: 10.1016/j.energy.2014.02.059.
[11] R. Chauhan, T. Singh, N. S. Thakur, and A. Patnaik, "Optimization of parameters in solar thermal collector provided with impinging air jets based upon preference selection index method," Renew. Energy, vol. 99, pp. 118–126, 2016, doi: 10.1016/j.renene.2016.06.046.
[12] M. Belusko, W. Saman, and F. Bruno, "Performance of jet impingement in unglazed air collectors," Sol. Energy, vol. 82, no. 5, pp. 389–398, 2008, doi: 10.1016/j.solener.2007.10.005.
[13] M. Zukowski, "Heat transfer performance of a confined single slot jet of air impinging on a flat surface," Int. J. Heat Mass Transfer, vol. 57, no. 2. pp. 484–490, 2013, doi: 10.1016/j.ijheatmasstransfer.2012.10.069.

[14] E. I. Esposito, "Jet impingement cooling configurations for gas turbine combustion," pp. 46, 2006.
[15] S. Yadav and R. P. Saini, "Numerical investigation on the performance of a solar air heater using jet impingement with absorber plate," Sol. Energy, vol. 208, pp. 236–248, August 2020, doi: 10.1016/j.solener.2020.07.088.
[16] S. Yadav and R. P. Saini, "Comparative study of simple and impinging jet solar air heater using CFD analysis," AIP Conf. Proc., vol. 2273, November 2020 pp. 1–8, doi: 10.1063/5.0024241.
[17] S. Yadav and R. P. Saini, "Thermo-hydraulic CFD analysis of impinging jet solar air heater with different jet geometries," in Lecture Notes in Mechanical Engineering, R. Kumar, A. K. Pandey, R. K. Sharma, and G. Norkey, Eds. Berlin, Germany: Springer Nature Publishing, 2021, pp. 193–201.
[18] A. Kumar, V. Ranjan Kaushik, and A. kumar, "A CFD approach to show the performance of solar air heaters with corrugated absorber plate," Journal for Advanced Research in Applied Sciences. vol. 4, no. 11, 2017, pp. 46–66.
[19] W. Gao, W. Lin, T. Liu, and C. Xia, "Analytical and experimental studies on the thermal performance of cross-corrugated and flat-plate solar air heaters," Appl. Energy, vol. 84, no. 4, pp. 425–441, 2007, doi: 10.1016/j.apenergy.2006.02.005.
[20] A. A. El-Sebaii, S. Aboul-Enein, M. R. I. Ramadan, S. M. Shalaby, and B. M. Moharram, "Investigation of thermal performance of-double pass-flat and v-corrugated plate solar air heaters," Energy, vol. 36, no. 2, pp. 1076–1086, 2011, doi: 10.1016/j.energy.2010.11.042.
[21] Y. Varol and H. F. Oztop, "A comparative numerical study on natural convection in inclined wavy and flat-plate solar collectors," Build. Environ., vol. 43, no. 9, pp. 1535–1544, 2008, doi: 10.1016/j.buildenv.2007.09.002.
[22] A. Karim and M. N. A. Hawlader, "Performance investigation of flat plate, v-corrugated and finned air collectors," vol. 31, pp. 452–470, 2006, doi: 10.1016/j.energy.2005.03.007.
[23] W. Gao, W. Lin, and E. Lu, "Numerical study on natural convection inside the channel between the flat-plate cover and sine-wave absorber of a cross-corrugated solar air heater," Energy Convers. Manag., vol. 41, no. 2, pp. 145–151, 2000, doi: 10.1016/S0196-8904(99)00098-9.
[24] ANSYS Fluent 14.0 User's Guide, 2011.
[25] D. S. Thakur, M. K. Khan, and M. Pathak, "Solar air heater with hyperbolic ribs: 3D simulation with experimental validation," Renew. Energy, vol. 113, pp. 357–368, 2017, doi: 10.1016/j.renene.2017.05.096.

[26] S. Polvongsri and F. O. F. Engineering, "Ashrae Standard 93-2003 methods of testing to determine flat-plate solar collectors by," pp. 5–9, 2013.
[27] S. Saedodin, S. A. H. Zamzamian, M. E. Nimvari, S. Wongwises, and H. J. Jouybari, "Performance evaluation of a flat-plate solar collector filled with porous metal foam: Experimental and numerical analysis," Energy Convers. Manag., vol. 153, pp. 278–287, August 2017, doi: 10.1016/j.enconman.2017.09.072.
[28] L. R. R. S. S. K. Deepak, "Thermal and fluid flow analysis of an corrugated and wavy channels: 'A comprehensive review,'" Int. J. Sci. Res., vol. 7, no. 1, pp. 165–169, 2018, doi: 10.21275/ART20179074.
[29] A. S. Yadav and J. L. Bhagoria, "A CFD based thermo-hydraulic performance analysis of an artificially roughened solar air heater having equilateral triangular sectioned rib roughness on the absorber plate," Int. J. Heat Mass Transfer, vol. 70, pp. 1016–1039, 2014, doi: 10.1016/j.ijheatmasstransfer.2013.11.074.
[30] V. B. Gawande, A. S. Dhoble, D. B. Zodpe, and S. G. Fale, "Thermal performance evaluation of solar air heater using combined square and equilateral triangular rib roughness," Aust. J. Mech. Eng., vol. 00, no. 00, pp. 1–11, 2018, doi: 10.1080/14484846.2018.1519986.
[31] A. Bakker, "Lecture 10 - turbulence models applied computational fluid dynamics," Lect. Notes, no. 2002, pp. 1–47, 2006.

Chapter 7

Leakage Current in Solar Photovoltaic Modules

Ravi Kumar[1], Manish Kumar[2], and Rajesh Gupta[1]

[1]Department of Energy Science and Engineering, Indian Institute of Technology Bombay, India
[2]Institute for Energy Technology Kjeller, Norway
Corresponding author: ravikr@iitb.ac.in, naik.manish17@gmail.com

Abstract

A photovoltaic (PV) cell is a semiconductor device which converts light energy into electricity. A large number of cells comprise a PV module. In a PV system, these modules are connected in series and parallel arrays divided into different strings. This arrangement of modules develops high voltage stress on a solar cell as modules are always grounded for safety reasons. A current is generated under this voltage stress, known as leakage current. Along with this leakage current, the availability of an adequate number of ions (i.e., Na^+) on the solar cell surface leads to potential induced degradation (PID). This results in the degradation in the performance of a solar cell. Therefore, leakage current can be used as a deterministic parameter for PID. There are different paths available for leakage current to flow. This leakage current depends on many factors, which can be categorized as module components and environmental conditions. Temperature and humidity are important factors in deciding the amount of leakage current.

7.1 Introduction

A photovoltaic (PV) cell is a semiconductor device that converts sunlight into electricity. It composes of a thin silicon wafer with n-type and p-type doped layers connecting with metal contacts on both sides, which collect current from the generation sites, as shown in Figure 7.1. It works on the principle

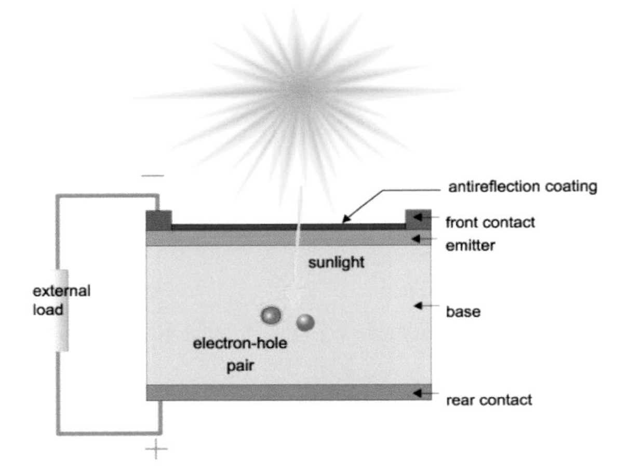

Figure 7.1 Cross-sectional view of solar cell [1].

of the photovoltaic effect. Let us see the working of solar cells under illuminated conditions in detail.

There are three important phenomena that take place inside the solar cell when sunlight hits on its surface, which are as follows:

i. **Excitation of free charge carriers due to sunlight absorption:** When sunlight strikes on the surface of the solar cell, the free charge carriers, i.e., electrons and holes, get excited by absorbing energy from the sunlight.

ii. **Separation of charge carriers as electron and hole:** Later, these charge carriers are separated due to the electric field at the p-n junction. Here, separation means these charge carriers move in an opposite direction from each other.

iii. **Collection of charge carriers by metal contacts:** At the last step, both the charge carriers (electrons and holes) are collected by metal contacts attached to both sides of the outer surface of the silicon wafer. This collection of charge carriers makes the metal contact polar in nature. Therefore, the metal contact where electrons are collected behaves as a negative terminal, while the other, where holes are collected, behaves as a positive terminal.

In short, we can say that a solar cell is a p-n junction semiconductor device, and this p-n junction inside the solar cell results in an electric field. When sunlight hits the solar cell's surface, electrons spring up and attract toward

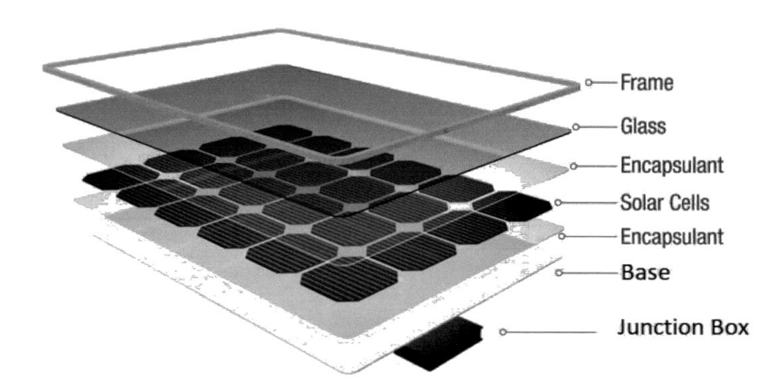

Figure 7.2 Cross-sectional view of PV module [2].

an n-type semiconductor. At the same time, the paired holes created by these electrons' vacancies move toward a p-type semiconductor where the electron combines with a hole after traveling through an external circuit that generates an electric current.

Solar cells always operate under the sunlight in open areas; therefore, in order to prevent these cells from harsh environmental conditions, i.e., humidity, dew, rain, and high temperature, these cells are encapsulated with additional materials such as glass and back-sheet along with ethyl vinyl acetate (EVA). These encapsulated constructions having solar cells in between are solar PV modules, as shown in Figure 7.2.

Temperature and humidity are important factors determining semiconductor devices' performance and reliability. The same thing applies to PV cells as well. The material used as a protective glass, i.e., soda lime glass or quartz glass, should be resistant to temperature changes and UV rays' exposure. The EVA used as encapsulant material prevents the moisture ingress into the PV module. Glass also resists mechanical stress up to an optimum level and provides strength to the PV module. An aluminum frame is attached to the structure in order to make the structure strong, sturdy, and easy to mountable.

A single module is not enough to generate enough power as many appliances require; hence, many modules are needed to connect in series and parallel arrays. These modules array along with some other essential devices, i.e., an inverter is known as a PV system. In a PV system, the aluminum frames of modules are always grounded for safety purposes. Numerous modules connected in a string and their grounding cause huge potential stress to solar cells of PV modules, as shown in Figure 7.3(a).

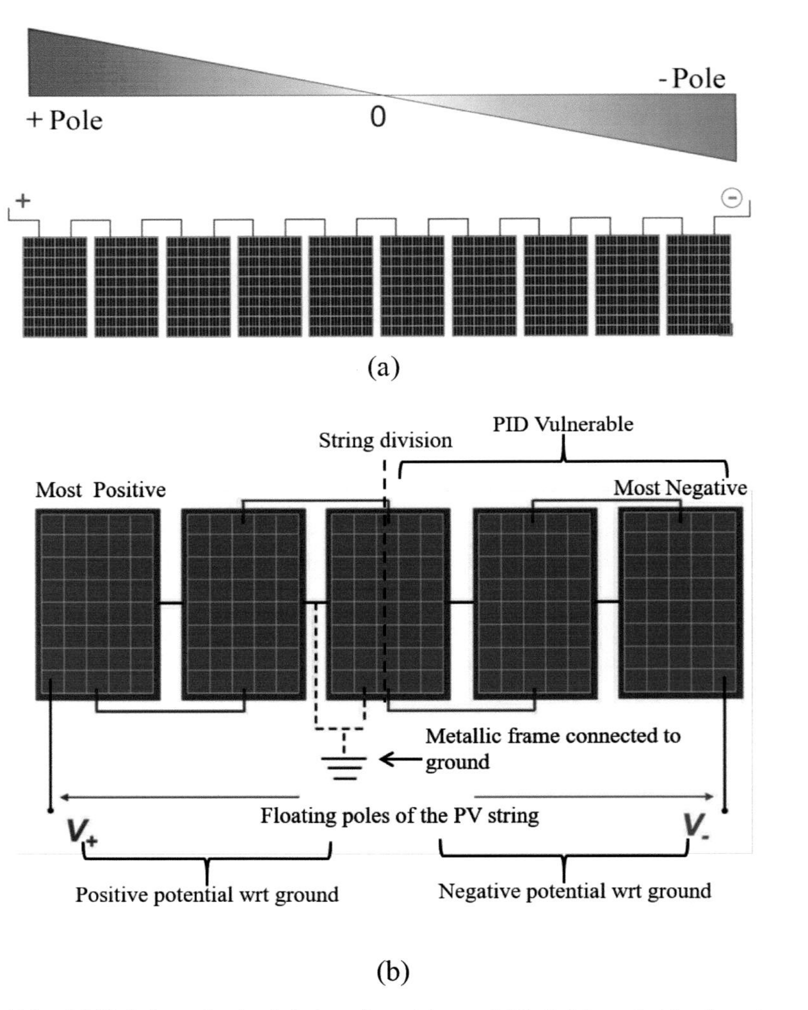

Figure 7.3 (a) Variation of potential stress in a string and (b) division of string based on cell potential with respect to ground.

7.2 Origin of Leakage Current

There are different types of topologies at the system level, i.e., with a transformer or without a transformer. For a transformer-less system in which neither the negative nor positive pole is grounded, a string is divided into two parts based on the potential difference between a cell with respect to the ground, which is depicted in Figure 7.3(b). Half of the string bears negative

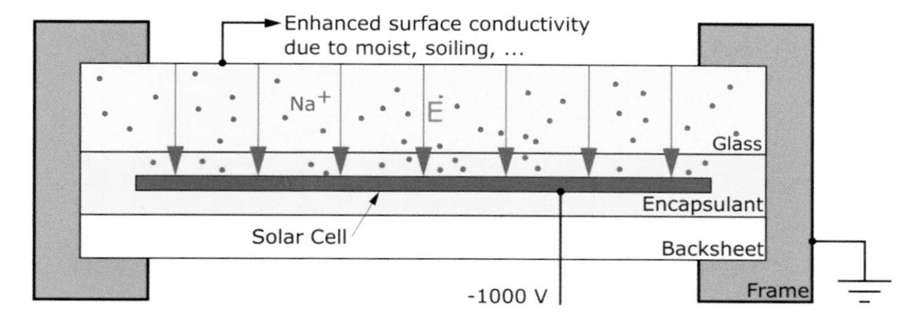

Figure 7.4 PID phenomenon in a solar cell [3].

voltage stress, while the other half bears positive voltage stress, which builds up due to a series of connections of modules in a string.

Under this high voltage stress, a current which is known as leakage current flows between the solar cell and module frame. This leakage current results in a reduction in the power output of a cell. The magnitude of this high voltage stress leakage current can be considered as an indicator of the susceptibility toward potential induced degradation (PID). It is a kind of degradation in which shunt formation takes place across the p-n junction as a result of an accumulation of ions (i.e., Na+) as shown in Figure 7.4. These ions drift under high voltage stress along with leakage current. In this case, poles in a string are always floating. The half part of the negatively stressed string becomes vulnerable to PID because negative potential on the surface of a cell attracts positive ions toward itself.

7.3 Paths of Leakage Current

Different paths are available for the leakage current to flow from the cell to the module frame. Some possible pathways for leakage current to flow in a PV module, as shown in Figure 7.5, can be described as follows:

i. Across the encapsulant and through the bulk of glass.

ii. Across the encapsulant and through the interface of glass and encapsulant.

iii. Through the bulk of encapsulant.

iv. Across the encapsulant and through the interface of encapsulant and back-sheet.

v. Across the glass and encapsulant and through the surface of the glass.

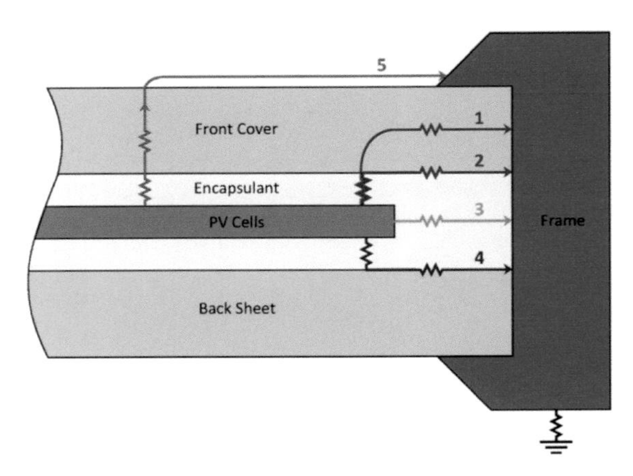

Figure 7.5 Different pathways through the components of PV module for leakage current [4].

From Figure 7.5, it can be observed that the total magnitude of the leakage current is a parallel combination of all the current from different pathways. This combination makes it difficult to identify the real culprit path responsible for degradation. Therefore, the need arises to measure the leakage current from different paths.

On the same line, different models have been proposed based on the finite element method to calculate the leakage current contribution through different paths of a module [5]. Based on these methods, some of the shares of different paths are shown in Figure 7.6. Analysis based on total leakage current does not provide insight into the specific material which is responsible for degradation. This detailed analysis of leakage current paths needs to be studied under high voltage stress. These dominant paths are being identified to quantify the physical and chemical changes occurring within the packaging materials. Based on this, the more vulnerable part to degradation of a module is identified.

7.4 Impact of Different Factors on Leakage Current

The PV module leakage current was studied first time by the Jet Propulsion Laboratory (JPL) in the early 1980s. This leakage current depends on various factors. These factors decide the path of leakage current under different conditions, i.e., at high humidity and wet conditions, leakage current through soda-lime silicate glass dominates while the dominant path is along with the EVA/glass interface [7].

There are two main factors on which leakage current depends:

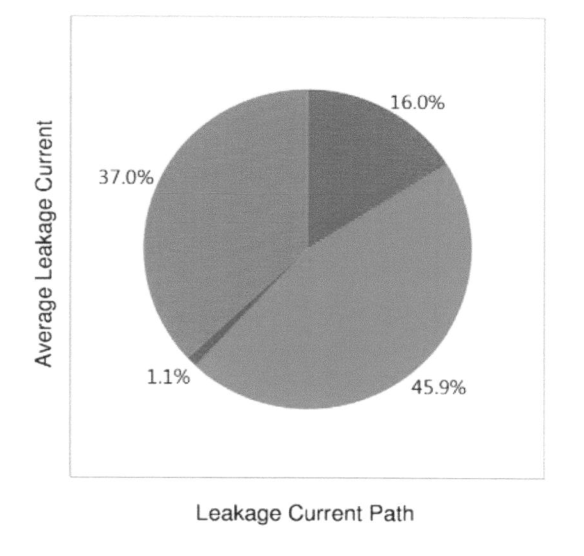

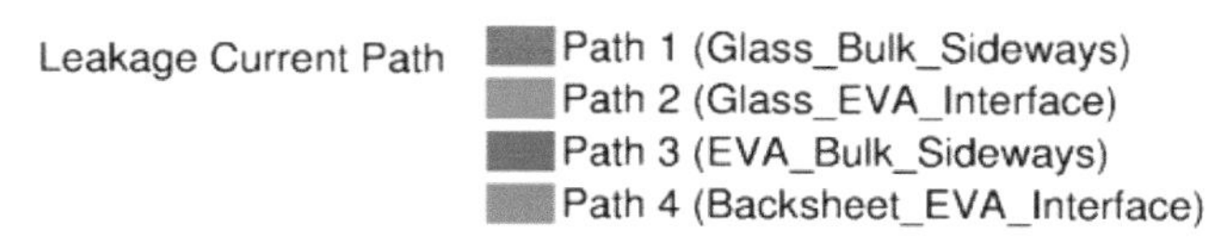

Figure 7.6 Shares of different pathways in total leakage current [4].

i. **Module factors** *(glass surface resistance, glass composition, and encapsulant resistance)*
ii. **System/environmental factors** *(module surface temperature, humidity, rain, dust, dew, string voltage, etc.)*

7.4.1 Modeling of leakage current

A model has been established for leakage current flowing through modules under high voltage stress conditions at given humidity and temperature [7]. The model given is as follows:

$$i(RH,V_b,T) = i_0(RH,V_b)\exp\left(-\frac{E_A(RH)}{kT}\right) \tag{7.1}$$

where i is the leakage current, E_A is the activation energy, RH is humidity, V_b is stress voltage, k is Boltzmann constant, and T is temperature.

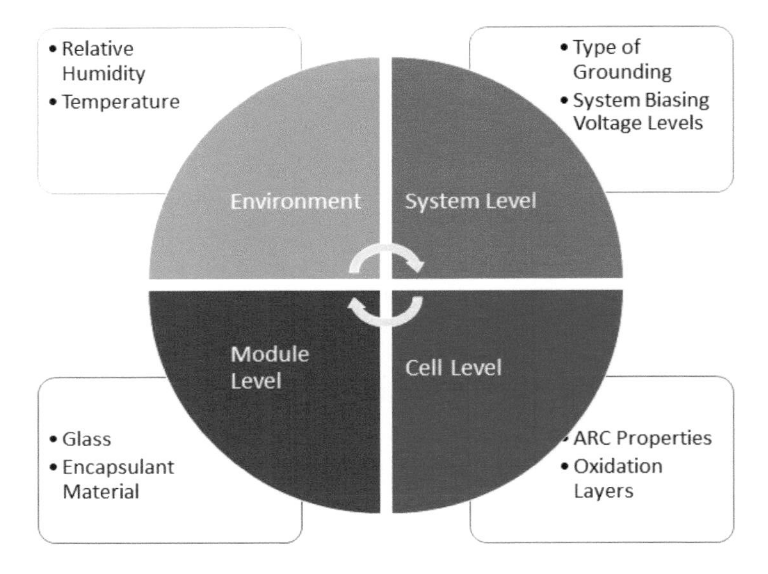

Figure 7.7 Influencing factors of leakage current.

The above model shows that leakage current is a function of humidity, stress voltage, and temperature. Here, the pre-factor is a function of humidity and stress voltage (Figure 7.7).

7.4.2 Module factors and their influence

a) Glass material: In order to make a solar cell durable, many packaging is used. In these packaging materials, glass is one which is used to cover solar cells. Soda-lime glass is used because of its low cost and good physical properties. However, the glass used here makes conditions favorable for PID. These glasses contain a significant amount of alkali metal ions (predominantly Na+) and have only moderate bulk resistivity. These conditions result in a large leakage current under high voltage stress.

b) Encapsulant materials: Encapsulant is used for thermal stability, resistance to moisture resistance, stability against UV radiation, and electrical protection for the module components. A variety of encapsulant materials are available in the market. Generally, ethyl vinyl acetate (EVA) is used as an encapsulant material. The conductivity of encapsulant is favorable for PID because it provides a smooth path for ions to drift under high-stress conditions, which constitutes leakage current.

7.4.3 Environmental factors and their influence

The effect of different environmental factors can be studied in two conditions, i.e., under fixed voltage and under varying voltage. The main important factors are temperature and humidity. Now let us understand the impact of temperature on leakage current under fixed voltage as well as under varying voltage.

7.4.3.1 Module temperature

i) Under fixed voltage: The Arrhenius plot of leakage current at fixed voltage stress is shown in Figure 7.8. From the leakage current model, it can be observed that leakage current shows an exponential relationship with the inverse of temperature. Hence, as temperature increases, leakage current will also increase. It might be due to an incline in the conductivity of module components with temperature. Figure 7.8 shows a similar slope for all three different kinds of situations defined as:

a) the module at 85 % RH with aluminum frame grounded;

b) module with aluminum cover on the glass surface;

c) the module under dry air ambiance.

It revealed that the different methods of connection have the same kinetics supported by the fact that the value of activation energy calculated using Arrhenius law is the same for all three cases, i.e., 75 kJ [8].

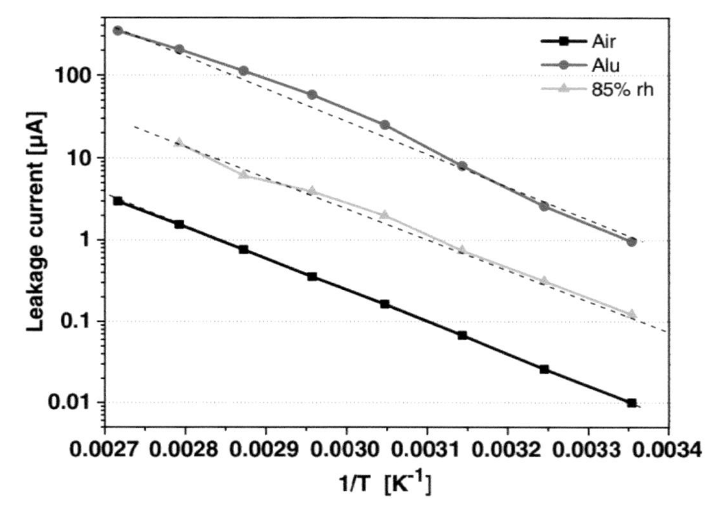

Figure 7.8 Arrhenius plot of leakage current with temperature [8].

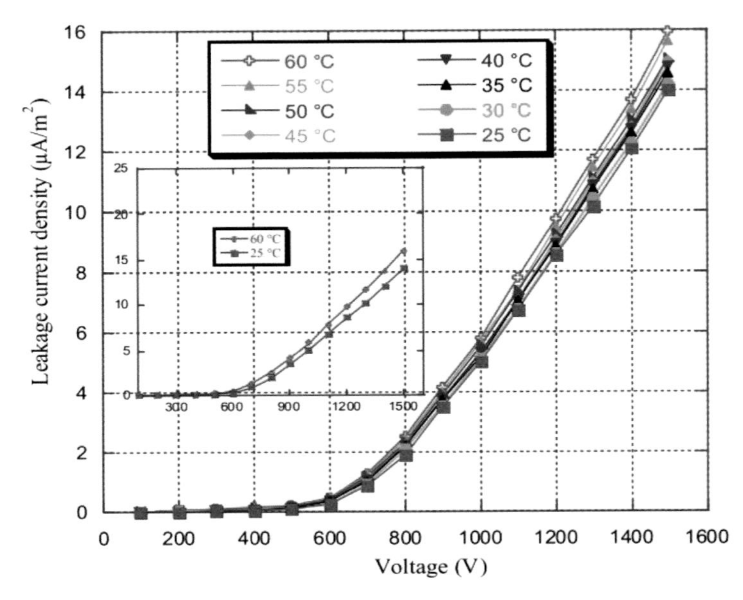

Figure 7.9 High voltage stress leakage current at different module temperatures [6].

ii) Under varying voltage. Here, Figure 7.9 depicts the leakage current variation of a PV module under different module temperatures with different high voltage stress. As it can be observed that humidity also changes with temperature, humidity also decreases as the temperature increases. Figure 7.9 shows the exponential increase in leakage current with stress voltage, but there is a minor increase in leakage current with the increase in module temperature. A slight increase in leakage current is the due change in material properties of glass and encapsulant materials. Glass and encapsulant material resistivity decreases at higher temperatures (soda-lime glass resistivity decreases by 0.002 $\Omega°C^{-1}$) [8]. The rate of increase in leakage current also increases with voltage stress. Some reported rate values from Figure 7.9 are 0.0067 $\mu A\ m^{-2}\ °C^{-1}$ at 600 V, 0.0222 $\mu A\ m^{-2}\ °C^{-1}$ at 1000 V, and 0.0524 $\mu A\ m^{-2}\ °C^{-1}$ at 1500 V [6].

7.4.3.2 Humidity or wet surface condition

i) Under fixed voltage: Humidity is necessary for the leakage current to flow. Figure 7.10 shows leakage current relation with humidity at a fixed voltage. From this, it can be observed that the leakage current is almost zero as humidity is below 50%. When it is crossed with a humidity value of about 60%, maximum leakage current flows under a certain value of high voltage

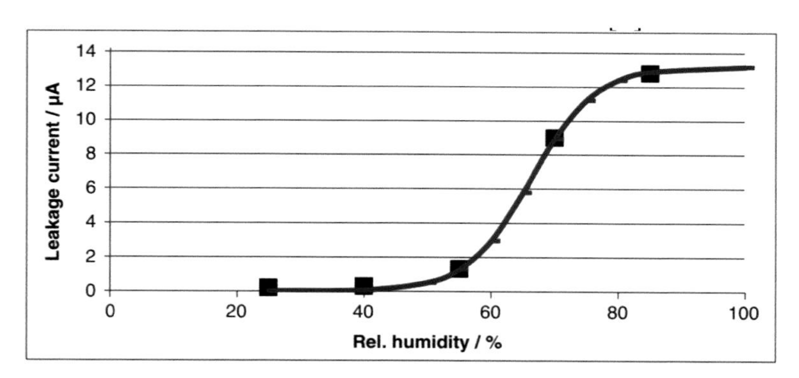

Figure 7.10 Impact of humidity on leakage current under constant high voltage stress [8].

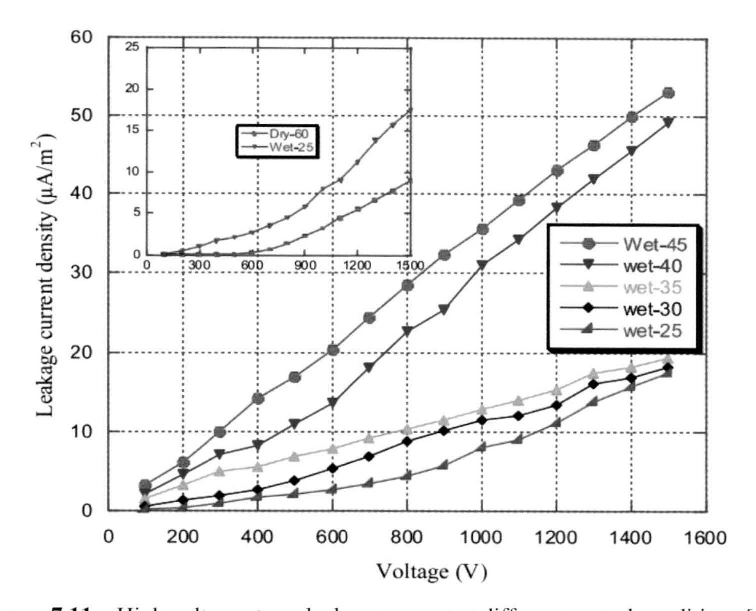

Figure 7.11 High voltage stress leakage current at different wetted conditions [6].

stress. This behavior can be addressed based on the water film that becomes electrically conductive after a threshold of about 60% RH [8].

ii) Under varying voltage: There are many reasons for the wetness of a PV module, i.e., high RH, dew, rain, etc. Figure 7.11 depicts the leakage current variation of wetted modules at different temperatures with high voltage stress. From Figure 7.11, it is observed that the wet leakage current of a module increases with temperature. The inset of Figure 7.11 shows that the wetted leakage current at room temperature is more than the leakage current at 60°C under dry conditions. The generation of charge carriers increases with an increase in temperature on wet module. The rate of PID increases at high temperature and wet surface conditions because, in this situation, the high voltage stress leakage current increases drastically [6]. The increase in leakage current with temperature when the module is wet can be seen as the temperature increases the generation of charge carriers also increases. The rate of PID increases at high temperature and wet surface conditions because, in this situation, the high voltage stress leakage current increases drastically [6]. For a fixed wet condition, the increase in leakage current seems to be linear with voltage which follows the leakage current model.

7.5 Conclusion

In a PV system, a large number of modules are connected in series and parallel arrays. This arrangement of modules results in high voltage stress on the solar cell as the modules are grounded for safety reasons. Under high voltage stress, ions, i.e., Na^+ present on the surface of a solar cell, drift toward the cell along with leakage current and accumulate inside the solar cell. This accumulation of ions penetrates the p-n junction and establishes a metallic contact which behaves as a shunt. This shunting is known as potential induced degradation (PID), and it results in a large drop in the performance of a solar cell. It is happening along with leakage current; therefore, leakage current is used as a determinant for PID. There are different paths available for leakage current to flow. This different path-based analysis is used to determine the most susceptible component of the module for degradation. There are different factors that affect the leakage current. They are mainly divided into two parts: module components and environmental factors. For module components factors, glass and encapsulant materials play a crucial role as they determine the resistivity against the leakage current. In the case of environmental factors, temperature shows a non-linear relationship with leakage current, while humidity shows a some-what linear increase in leakage current. These environmental factors result in an increase in the conductivity of module components; thus, leakage current occurs.

References

[1] Web edition of PVCDROM [Accessed on: 12-10-2021]. https://www.pveducation.org/pvcdrom.
[2] Clean Energy Reviews [Accessed on: 13-10-2021]. https://www.cleanenergyreviews.info/blog/solar-panel-components-construction.

[3] Energy Ville [Accessed on: 12-10-2021]. https://www.energyville.be/en/press/expert-talk- potential-induced-degradation-photovoltaic-modules.

[4] Neelkanth G. Dhere, Narendra S. Shiradkar, and Eric Schneller. Device for detailed analysis of leakage current paths in photovoltaic modules under high voltage bias. *Applied Physics Letters*, 104(11):112103, 2014.

[5] Neelkanth G. Dhere, Narendra S. Shiradkar, and Eric Schneller. Device for comprehensive analysis of leakage current paths in photovoltaic module packaging materials. In *2014 IEEE 40th Photovoltaic Specialist Conference (PVSC)*, pages 2007–2010. IEEE, 2014.

[6] Mohammad Aminul Islam, Md Hasanuzzaman, and Nasrudin Abd Rahim. Effect of different factors on the leakage current behavior of silicon photovoltaic modules at high voltage stress. *IEEE Journal of Photovoltaics*, 8(5):1259–1265, 2018.

[7] J. A. Del Cueto and T. J. McMahon. Analysis of leakage currents in photovoltaic modules under high-voltage bias in the field. *Progress in Photovoltaics: Research and Applications*, 10(1):15–28, 2002.

[8] Stephan Hoffmann and Michael Koehl. Effect of humidity and temperature on the potential-induced degradation. *Progress in Photovoltaics: Research and Applications*, 22(2):173–179, 2014.

Chapter 8

Reliability and Degradation Analysis of Crystalline Silicon Photovoltaic Module

Roopmati Meena, Manish Kumar, and Rajesh Gupta

Department of Energy Science and Engineering, Indian Institute of Technology Bombay, India
Corresponding author: meenaroopmati@iitb.ac.in

Abstract

Long-term reliability of operation is essential for solar photovoltaic (PV) modules due to the widespread adaptation of solar PV technology as an alternative green source of energy. The reliability of the PV module broadly depends on the reliability of its various components when exposed to external environmental conditions. The main environmental factors that affect PV module components are extreme temperature conditions, moisture, ultraviolet (UV) radiation, high-speed wind, hailstorm, etc. At any location or site, the PV module is subjected to the combination of these external environmental factors simultaneously. As a result, PV modules suffer from different types of defects and degradations (D&D) during field operation. The origin of some of these D&D could be traced back to the manufacturing, transportation, and installation of solar cells and PV modules. However, D&D, which originate during the early stage of cells or module, could propagate into severe D&D in outdoor conditions.

Moreover, some of the D&D start originating within a few years of PV module operation and manifest into critical loss incurring modes. Thus, it is important to understand the operating mechanism of various environmental factors on the operation of PV modules separately and in tandem. In addition, understanding the mechanism responsible for D&D formation in various components paves the path for chemical and structural improvement for future PV technologies and module architecture to minimize or avoid power loss modes. In this regard, this chapter discusses the main factors responsible for D&D in PV modules. In addition, major types of D&D reported in the PV modules have also been discussed.

8.1 Introduction

Energy being the major driving force and an indication of the standard of living has become a necessity in day-to-day life. In the wake of ever-increasing carbon emissions and the threat of global warming, the need of the hour is to switch to renewable sources of energy. Among various renewable sources of energy, solar photovoltaics has shown promising results in recent times in combating the fight against power shortages, remote area connectivity, and, most importantly, reducing carbon emissions by providing green electricity with power decentralization as an added advantage. Moreover, the harnessing of replenishable sources of energy serves twofold benefits. First, it enables the preservation of the underground resources, which have taken centuries to become crude oil or gases, as an emergency backup. Second, it reduces the havoc of pollution created due to the uncapped use of fossil-based energy sources. In this regard, the harnessing of the sun's energy through photovoltaic (PV) modules has gained significant momentum worldwide in recent decades due to rapid advancement in technological know-how and widespread adoption due to government policy interventions and public awareness. The growth of the solar energy sector has shown promising trends in the past few years as an alternative energy source. The year 2019 witnessed an annual growth of 22% in power generation through solar PV. Wherein, total global energy generation by solar PV has been estimated to stand at 720 TWh in 2019 [1]. By the end of the year 2020, total solar PV installation worldwide stood at 7604 GW [2].

Rapid installation of PV modules has led to an increase in the production of PV modules, with China leading the manufacturing market. Among various PV technologies, crystalline silicon (c-Si) PV modules account for a major fraction of installed PV modules due to their low cost, manufacturing know-how, and availability of a significant research database. Large-scale manufacturing of modules can often be characterized by compromise in the quality of raw materials and integrity in the module structure. As a result, the PV module starts to have defects and degradation (D&D) from the early stage of installation. Figure 8.1 categorizes the main D&D observed in PV modules based on their expected occurrence during the life-span of the module. Infant-failure accounts for power loss modes in newly installed modules. These D&D are mainly caused during the PV modules' manufacturing, transportation, and installation. Midlife failures result from exposure of PV

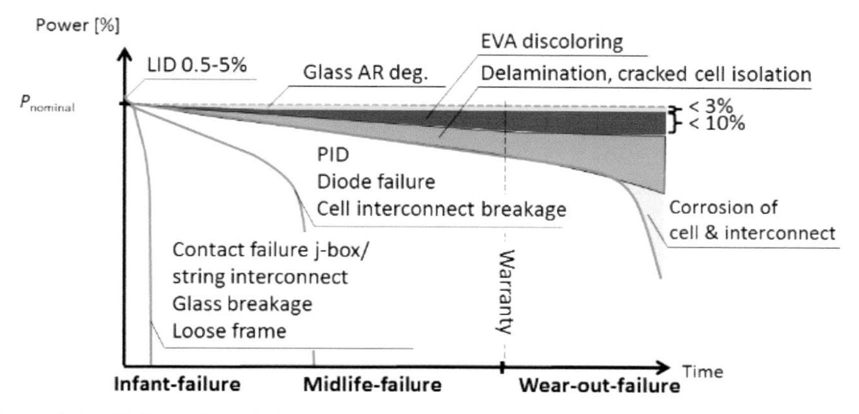

Figure 8.1 Various degradation and defect scenarios over the life-span of the c-Si PV module [3].

modules to harsh external environmental conditions of excess humidity, high temperature, ultraviolet (UV) radiation, etc. Wear-out failures determine the age of operation of the module. PV modules are decommissioned if the module power reaches below a certain level (70%–80%) concerning its initial power rating [3].

To achieve the target of sustainable development goals, the solar PV module installations need to last for a guaranteed 25 years. However, degradation of the various components of the PV module structure directly or indirectly affects the electrical power output. In addition, the presence of D&D also accelerates the deterioration of the PV module. This, in turn, reduces the lifetime of modules and return on investment. Hence, it becomes important to understand the underlying factors and operating mechanisms that lead to the failure of solar PV modules in the field operating conditions.

8.2 Factors Affecting the Reliability of PV Modules

Various components of PV modules suffer from structural, chemical, and optical D&D resulting in electrical and thermal losses. The PV module is vulnerable to D&D at various stages from its manufacturing to transportation to installation in the field. In addition, a PV module, when installed in an external environment, experiences various kinds of stressors such as high wind, humidity, high temperature, etc., and their variation under various climatic conditions throughout the year. As a result, these defects or degradation in the PV module cause a reduction of the output power. At the same time, it can also contribute to the acceleration of pre-existing defects or degradation, thus, affecting the long-term reliability of the operation of PV modules. In this section, various factors responsible for affecting the reliability of a PV module at different stages have been discussed in detail.

8.2.1 Manufacturing

Various factors at the manufacturing stage of solar cell and PV modules, such as impurities in raw material, mishandling of sensitive equipment, improper gridline printing, improper lamination parameters, poor soldering process, inferior quality of raw material, etc., contribute toward either formation or initiation of various types of defects and degradation. These defects and degradations could reduce the efficiency of the finished solar cell or could increase in severity at a later stage. Shunt formation is one of the most prominent defects originating during the manufacturing stage of silicon solar cells due to impurities in the raw material [4]. Shunting is the phenomenon of forming an alternative path for current to flow through the pn-junction of the solar cell, which causes light-generated current to sink through it, thus resulting in the power loss of the cell/module. Another type of defect which can originate during the manufacturing stage is the micro-cracks in the silicon wafer. Micro-cracks can be formed during the thin slicing of silicon wafers from the silicon ingot to form solar cells [5]. Micro-cracks can manifest themselves into severe cracks at later stages of module operation, which can result in significant power loss.

Moreover, during the printing of silver gridlines which are responsible for carrying light generated carriers in silicon cells, poor quality of silver paste or improper design of wire mesh could result in hindrance in the collection of light generated current [6]. Similarly, improper soldering of copper ribbon over the solar cell results in the residual thermal stress in the PV module, which increases under field operating conditions and can cause fracturing of solar cells [7]. During lamination of different layers of the PV module, incorrect temperature/lamination parameters or improper stacking of various layers could result in the formation of air bubbles between various layers, which could act as sites of heat accumulation in the PV module, thus reducing its performance.

8.2.2 Transportation and installation

During the transportation and installation of PV modules, they are subjected to various types of mechanical stressors [8]. During transportation, modules are stacked over each other. If sufficient cushioning is not provided between two stacked modules, the movement of the truck/vehicle on uneven roads and speed breakers will cause movement amongst modules. Such movements could result in micro-crack formation and propagation in the solar cells. In the worst-case scenario, it can also cause the breakage of the top glass in the module [3].

Similarly, improper handling of PV modules during installation in the field can also result in the formation of micro-cracks in the solar cell as well as propagation of pre-existing micro-cracks into severe cracks. PV modules are installed on the structures inclined at the latitude of a given location to intercept maximum incident radiation to maximize the output power. The improper alignment of structures supporting PV modules would induce permanent mechanical stress over the PV module structure. Moreover, high wind speeds in open terrain also exert mechanical pressure on the mounted PV module [9]. In the long run, the accumulation of such mechanical stress can cause permanent damage to the PV module.

8.2.3 External environmental conditions

The various external environmental factors also contribute to the formation of defects and degradation in the PV modules. These factors act slowly over the life-span of the module. Depending on the climate of a specific location, the prominent environmental factors and their intensity vary. As such, certain types of defects and degradations are more prominent in a specific climatic condition. The main factors available in external environments and their operating mechanism responsible for the reduction of PV module reliability are discussed in detail below.

8.2.3.1 High temperature and its cyclic variation

PV modules are tested for efficiency and output power under a standard temperature condition of 25°C. However, the PV module operates either at a higher temperature or a lower than standard temperature under all practical operating conditions. While at lower temperatures working efficiency of PV modules increases slightly, higher temperature conditions significantly reduce the operating efficiency of PV modules. The high temperature of the PV module increases the degradation rate by a factor of two for every 10°C rise in the module operating temperature [10]. Moreover, the PV module is a multi-layered structure made up of various types of materials. These materials behave differently under high-temperature conditions due to differences in their thermal properties. Thereby, thermal mismatch amongst various components in a PV module due to high temperature could result in D&D

initiation and propagation. In addition to high temperature, cyclic variation in the temperature during day and night also induce thermal stress on the PV module. Extremely high temperatures during day time and cold temperatures during the night can cause continuous expansion and contraction of components in the PV module. Such cyclic temperature variation could induce physical damage to the PV module [11]. Generally, high and extreme cyclic temperature conditions are mainly prevalent in desert climatic conditions where extreme temperature values can induce significant defects and degradation in PV modules. High-temperature conditions onset a closed-loop degradation mechanism wherein an increase in temperature results in degradation of the PV module, which again leads to an increase in temperature of the module.

8.2.3.2 Irradiance

A solar cell converts a section of the sun's spectrum (visible part of the spectra from 400 to 700 nm) into electricity while the remaining major portion does not get utilized. A solar cell is designed to operate at rated input irradiance of 1000 W/m^2. The amount of current generated in a solar cell is directly proportional to the input irradiance. Lower irradiance conditions reduce the output current due to the reduced generation of charge carriers. While an extremely higher dose of incident irradiance also reduces the output power as the proportion of unused energy increases. The unused energy gets converted into heat, which increases the module temperature. The high-temperature operation of the PV module reduces its efficiency and compensates for the increase in output power from the increase in light-generated current. Also, the low wavelength part of the sun's spectrum, from 10 to 400 nm, contains UV radiations. Exposure of the PV module to the high dose of UVB (280–315 nm) radiation induces chemical degradation in its organic components, such as encapsulant and back sheet (Tedlar) [12]. Sites with high altitudes where the atmosphere is thin to absorb UV radiation or dry atmosphere which are void of air particles to absorb UV radiation receive high doses of UV radiation and hence witness a higher rate of polymeric degradation in the PV modules.

8.2.3.3 Humidity

High moisture content in the surrounding atmosphere could damage various metallic components of the PV module. Under high humidity conditions, metallic contacts in the junction box at the back-side of the module get damaged due to corrosion. Moisture can ingress through the back-side of the PV module due to the breathable nature of Tedlar and encapsulant material [13].

The exposure of the copper ribbon interconnector to moisture content would cause its corrosion. Moreover, the presence of moisture at busbar-ribbon assembly also makes soldering material prone to degradation. The metallic particles present in soldering material react with moisture to oxidize into non-conductive material. In some cases, solder material has also been displaced from beneath the interconnect ribbon after exposure to high moisture conditions [14]. As a result, the contact between the interconnect ribbon and the silver busbar is affected. This results in a hindrance to the current flow between solar cells. In addition, moisture can travel laterally to the front side of solar cells from the white space present between solar cells. Here, moisture could react with silver finger gridlines at the cell edges, resulting in the corrosion of the finger [15]. It disrupts the collection of light-generated current from solar cells, thereby reducing the output power of the PV module. It has been observed that higher temperature conditions aid in moisture ingression into the PV module [16]. In general, coastal areas are characterized by high moisture content in the atmosphere. Moreover, floating PV module installations are also subjected to higher humidity [17–19].

8.2.3.4 Wind, sand, snow, and hailstorm

High-speed winds would cause vibrational motion in the mounted PV modules. These vibrational movements subject the module to bending stress which can cause micro-cracks in the solar cells. Regular exposure to similar high-speed wind conditions could result in the conversion of micro-cracks into severe cracks. Moreover, in desert conditions, the high-speed wind is generally accompanied by coarse sand particles and dust [20]. Accumulation of sand or dust particles over front-side glass reduced the incident solar radiation reaching the solar cell, thus reducing output electrical power [21]. The movement of sand particles over the front glass of the PV module causes its abrasion and the formation of scratch marks. In addition, high-speed snow and hailstorm act as impact forces. It can cause the breakage of glass and the formation of cell cracks in the PV module [22].

8.3 Different Types of Defects and Degradation in PV Modules

A PV module installed in outdoor environmental conditions is acted upon by various factors simultaneously. It leads to the formation of multiple types of defects and degradations in various module components. Major defects and degradation observed in various components of the PV module are discussed in detail in the following sections.

8.3.1 Encapsulant degradation

Encapsulant is a thin transparent layer of organic material that serves multiple purposes in a PV module. The main purpose of the encapsulant is to provide protection to cell metallization and interconnect ribbon from moisture ingression as well as electrical insulation. It also provides adhesion between the different layers of the module. Encapsulant also acts as a shock-absorbing layer, protecting fragile solar cells from impact shocks [23]. Many materials are available as an encapsulant for the PV module, such as silicones, Ionomers, polyolefin elastomers, and ethylene-vinyl acetate (EVA). EVA is the most widely used commercial encapsulant material due to its economic viability [24]. EVA is an organic copolymer of ethylene and vinyl acetate. EVA used in PV modules is incorporated with many additives to protect it from external environmental conditions. However, EVA degrades when subjected to high temperature, humidity, and UV radiation. Two major degradation modes of EVA are discoloration and delamination, which are discussed in detail below.

8.3.1.1 Discoloration

Discoloration of the EVA is one of the major degradation modes observed in the field-aged PV modules. Under high temperature and UV irradiance, EVA becoming yellow or brown from transparent is termed discoloration, as shown in Figure 8.2. It can be attributed to the decomposition of the main chain of EVA by high-energy UV radiation and high temperature [25]. The by-products formed due to the degradation of EVA impart a yellow or brown color to the EVA. The reaction involving discoloration releases acetic acid as a by-product, which causes corrosion of metallic components of the module, i.e., finger gridlines and interconnect ribbon [26, 27]. Discoloration generally

(a) (b)

Figure 8.2 Discoloration of encapsulant in field-aged PV module [29].

has a characteristic circular pattern over the center of cells. However, around the edges of the solar cell, moisture ingression reverses the process of discoloration. This phenomenon is known as photo-bleaching, wherein brown or yellow EVA again becomes transparent [23]. Discoloration has primarily been observed in the desert region, which has a significantly high amount of UV radiation and high-temperature conditions [28].

8.3.1.2 Delamination

Delamination is another type of encapsulant degradation prevalent in outdoor conditions. It weakens of adhesion strength between solar cell and encapsulant, leading to the formation of a void between two layers. Delamination can occur due to multiple reasons. Poor lamination process during the manufacturing stage resulting in the insufficient flow of EVA under interconnect ribbon tabbing could leave voids between the layers. Also, the chemical reactions involving the degradation of different components of the module release gases, which get trapped between the cell and the EVA layer. The trapped gases under high-temperature conditions could expand, resulting in the widening of the delaminated region [30].

Moreover, the additives added in the EVA to strengthen the adhesive bonds are reported to degrade under hydrothermal conditions. When these additives deplete, the interface becomes prone to degradation under external environment conditions [31]. Also, delamination has been observed to originate along busbar regions generally. The delaminated region facilitates the ingression of moisture and air through voids, resulting in corrosion of the metallic parts in the corresponding regions [32]. Delamination at the cell–EVA interface has a characteristic gray appearance, as shown in Figure 8.3.

Figure 8.3 Delamination of encapsulant along busbar in the field-aged PV module [30].

8.3.2 Metallization degradation

Silver gridline metallization (finger and busbar) and copper ribbon interconnect constitute the metallization component of the PV module. These are responsible for collecting light-generated current from the solar cell to the external circuit. These metallic components of PV modules are susceptible to corrosion degradation through an oxidation mechanism when exposed to moisture or oxygen [33]. When coming in contact with moisture or air, silver finger gridlines oxidize into the brown-colored compound of silver oxide, as shown in Figure 8.4(c). Similarly, copper interconnect ribbon also degrades into greenish-blue or brown components due to the formation of copper carbonate, as shown in Figure 8.4(d). The deposition of brown or greenish compounds over fingers and busbars changes the material's conductive properties [14, 34]. As a result, the flow of current gets interrupted at the location having the finger or busbar corrosion, resulting in loss of output power. Interconnect

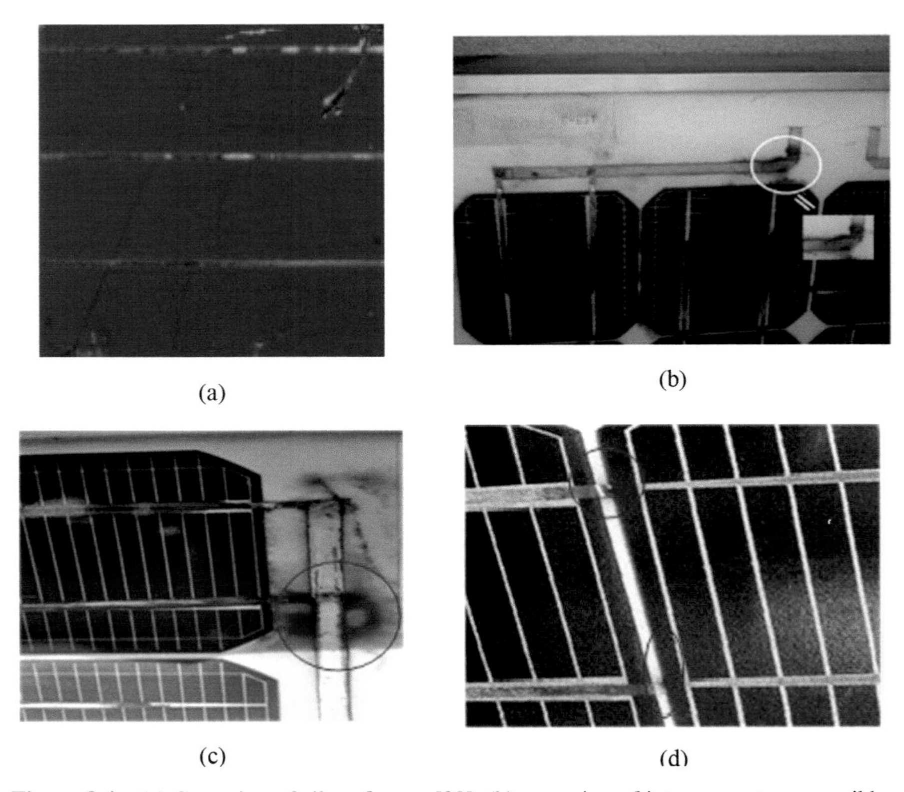

Figure 8.4 (a) Corrosion of silver fingers [38], (b) corrosion of interconnect copper ribbon [29], and (c) and (d) burnout of interconnect copper ribbon [37].

ribbon/wires are also found to burn out during the operation of the PV module in the field, as shown in Figure 8.4(a) and (b). It occurs due to a faulty soldering process which leaves space between two wires at the joints [35, 36]. A sudden surge in system voltage can result in the arc formation at a weak joint location, causing interconnect wire burning.

Moreover, the difference in the coefficient of thermal expansion between the metallization and other components of the PV module results in the development of thermo-mechanical stresses. Such stress can cause the breakage of these components. In addition, the contact of interconnect ribbon with silver busbar gets affected due to the degradation of solder material. The degradation of fingers and interconnect ribbons is one of the prominent degradation types seen in field-aged modules [37].

8.3.3 Shunt

The shunt formation in the solar cell mainly happens during the manufacturing stage. A shunt reduces the output current by providing an alternative path for light-generated current to sink. Shunting can occur at different stages of manufacturing due to various reasons. The presence of impurities during the formation of the pn-junction makes a conductive path across the junction. When printing silver paste to make fingers, if silver paste penetrates through the length of the emitter layer, it establishes direct contact with the base layer, providing a low resistance path for current to sink.

Similarly, improper edge isolation results in the formation of shunts [39]. The severity of the shunt depends on its location and proximity to current collection regions. The shunt at the edges or between fingers allows limited sinking of current from the concerned region; hence, the effect of the shunt is limited [40]. However, a shunt under the finger and busbar would cause the sinking of the module current. The severity of shunt under low irradiance conditions is higher. Under low irradiance conditions, the light-generated current is limited by incident irradiance; as a result, the proportion of current sinking in the shunt becomes comparable with light-generated current [41].

8.3.4 Cracks

Solar cells, micron-level thin, are fragile structures prone to cracking. Cracks in solar cells are a commonly observed defect in the field-aged PV modules. Based on the visibility of cracks to bare eyes, they are classified as micro-cracks and cracks, as shown in Figure 8.5(a) and (b). While the former is not visible to the bare eye, the latter can be observed as a broken cell area with an

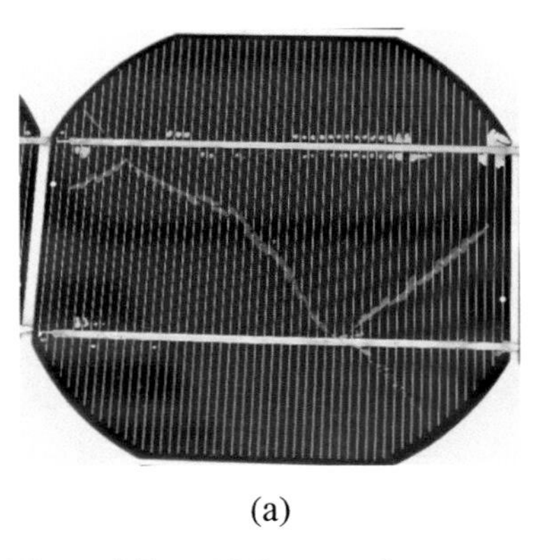

(a)

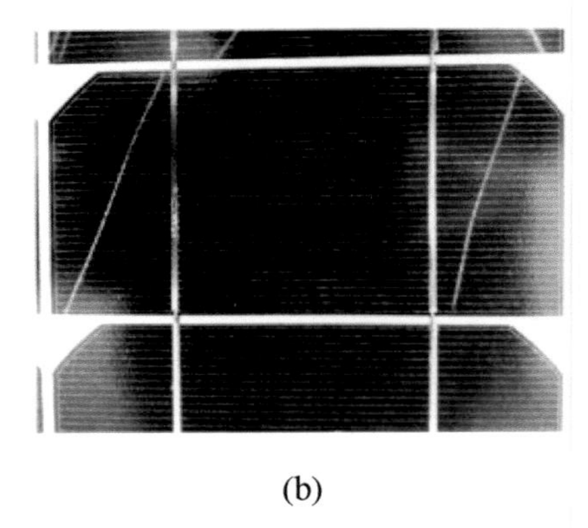

(b)

Figure 8.5 (a) Micro-crack accompanied with delamination along its length. (b) Crack with loss in the active area of the cell [44].

inactive loss. Cracks in the cells can originate at all stages of the PV module life-span due to various types of mechanical or thermo-mechanical stress [42].

Moreover, micro-cracks formed at any stage in the PV module can propagate into broken cell cracks under field operating conditions [43]. The regions have cracks, and micro-cracks act as favorable sites for moisture and air ingression, resulting in metallization corrosion around the cracked region. In some PV modules, the regions around micro-cracks are found to have a characteristic signature gray appearance of EVA delamination, as shown in Figure 8.5(a). This can be attributed to the degradation of encapsulants due to moisture ingression along the cracked region. The severity of cracks depends on the orientation of the crack with respect to the busbar and the loss of active area in the solar cell. The cracks that are parallel to the busbar are the most common type of cracks observed in the aged PV modules [8].

8.3.5 Potential induced degradation (PID)

In the recent decade, large-scale PV field installations have detected potential induced degradation (PID) of the PV module. Under working conditions of PV modules in a string, the system voltage reaches up to a few hundred volts. Depending on the grounding/earthing conditions of the string, this voltage would appear between the module frame and solar cells, causing the flow of leakage current through the structure of the module. In a p-type crystalline silicon solar cell, sodium (Na^+) ions present in the front glass migrate toward the cell via the encapsulant layer. These Na^+ ions penetrate through the emitter layer to the pn-junction, leading to shunt formation [45]. Unlike shunt

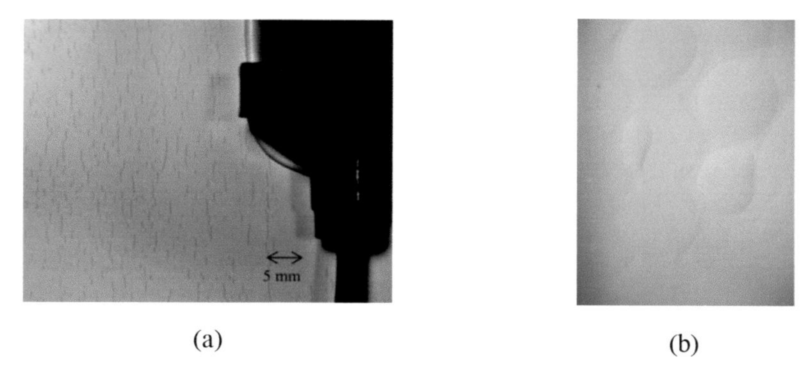

(a) (b)

Figure 8.6 (a) Cracking [49] and (b) bubble formation at the back-sheet layer [51].

formation due to impurities, PID-shunt affects a large localized area within the cells. The stalking of Na^+ near metallization has been found to initiate corrosion reactions in the presence of moisture and air [46]. The chemical reaction results in the release of gases, which leads to the delamination of EVA. High-humidity and high-temperature conditions provide a conducive environment for leakage of current through the module, thus favoring PID-induced losses [47]. The severity of cracks depends on the orientation of the crack with respect to the busbar and the loss of active area in the solar cell (Figure 8.6).

8.3.6 Back-sheet degradation

Back sheet is a multi-layer structure popularly used in conventional PV module construction at the back-side of the module. Recently, the back sheet has been replaced with glass in new module architectures. The main function of the back sheet is to provide electrical insulation to the module and protection from external environmental factors such as heat, humidity, UV radiation, and mechanical stability to the module. There are many types of back sheet available, namely, polyethylene terephthalate (PET), polyvinyl fluoride (PVF), polyvinylidene fluoride (PVDF), and polyamide (PA). PET-based back sheet are popularly used in commercial PV modules [48]. The back sheet is exposed directly to the external environment conditions, degrading under the influence of moisture, UV, and high temperature. The back sheet is discolored and cracked and has bubble formation [49]. The cracking of PET is the most commonly observed degradation of the back sheet in the field. Cracks in the back-sheet layer compromise the structural integrity of the layer and allow bulk ingression of moisture and air through it, resulting in degradation of metallization [50].

Moreover, delamination and bubble-like formation have been prominently reported as degradation in the back sheet. Such degradation could occur due to the accumulation of gases released during the chemical degradation of EVA or metallization at the back-sheet layer. Delamination of the back sheet restricts the heat flow through the module by trapping the heat in the air void. Degradation of the back sheet accelerates the degradation of other module components, thus affecting the overall reliability of the PV module.

8.4 Conclusion

Solar photovoltaic (PV) has become a well-established renewable source of power, which has started gaining significant momentum as an alternative source of power worldwide. With the life-span guarantee of 25–30 years for the PV modules operation, large-scale power plants of MW capacity have cropped up in the recent decades. In addition, various PV technologies and PV module architecture have also become commercially available. However, conventional crystalline silicon (c-Si) PV modules are still commercially popular due to their low cost, manufacturing know-how, and significant research databases. Yet, c-Si PV modules installed outdoors are susceptible to damage due to defect formation and degradation. Defects and degradations in the PV module pose a serious threat to its reliable operation. These defects and degradation restrict the power generation by the solar cell directly or indirectly by either blocking the incoming irradiance or by interrupting the collection of light-generated current.

Major modes of degradation and defect in a PV module are found to originate under harsh field operating conditions. Wherein, environmental factors such as high temperature, ultraviolet (UV) radiation, high moisture content, high-speed wind, etc., act simultaneously upon various components of the PV module, which leads to their degradation and defect formation. Degradation of encapsulant (EVA) in the presence of UV radiation and high temperature can be attributed as one of the major modes of degradation in PV modules. This is followed by degradation of metallization due to their corrosion facilitated by air and moisture ingression. Further, cracking in the back-sheet layer of the PV module provides a preferential pathway for moisture and air ingression.

Moreover, pre-existing defects such as micro-cracks propagate into major cracks under thermal stress conditions caused by cyclic variation in temperature day and night. In the recent decade, new modes of degradation such as potential induced degradation (PID) have also been discovered and are in the investigation phase. With the advancement in novel investigation techniques and methodologies, the possibility of fast detection and mitigation of defects and degradation in the modules has improved. However, in-depth analysis and investigation of existing defects and degradation in PV modules are of utmost importance for further improvement of existing technology as well as the creation of a knowledge base for new technologies.

8.5 Acknowledgment

The authors would like to thank the Indian Institute of Technology Bombay, Mumbai, India, for providing the resources to write this chapter.

References

[1] H. Bahar, Tracking Report-June 2020-Solar Energy, 2020. https://www.iea.org/reports/solar-pv (accessed June 29, 2021).
[2] IEA, Snapshot of Global P.V. Markets 2021, 2020. http://www.iea-pvps.org/fileadmin/dam/public/report/technical/PVPS_report_-_A_Snapshot_of_Global_PV_-_1992-2014.pdf.
[3] M. Köntges, S. Kurtz, C.E. Packard, U. Jahn, K. Berger, K. Kato, T. Friesen, H. Liu, M. Van Iseghem, Performance and reliability of photovoltaic systems subtask 3.2: Review of failures of photovoltaic modules: IEA PVPS task 13: external final report IEA-PVPS, 2014.
[4] O. Breitenstein, J. Bauer, J.P. Rakotoniaina, Material-induced shunts in multicrystalline silicon solar cells, Semiconductors. 41 (2007) 440–443. https://doi.org/10.1134/s106378260704015x.
[5] G. Dongre, R. Singh, S.S. Joshi, Response surface analysis of slicing of silicon ingots with focus on photovoltaic application, Mach. Sci. Technol. 16 (2012) 624–652. https://doi.org/10.1080/10910344.2012.731952.
[6] R. De Rose, A. Malomo, P. Magnone, F. Crupi, G. Cellere, M. Martire, D. Tonini, E. Sangiorgi, A methodology to account for the finger interruptions in solar cell performance, Microelectron. Reliab. 52 (2012) 2500–2503. https://doi.org/10.1016/j.microrel.2012.07.014.
[7] M. Mathusuthanan, K.R. Narayanan, K. Jayabal, In-plane residual stress map for solar p.v. module: A unified approach accounting the manufacturing process, IEEE J. Photovoltaics. 11 (2021) 150–157. https://doi.org/10.1109/JPHOTOV.2020.3029226.
[8] S. Kajari-Schröder, I. Kunze, U. Eitner, M. Köntges, Spatial and orientational distribution of cracks in crystalline photovoltaic modules

generated by mechanical load tests, Sol. Energy Mater. Sol. Cells. 95 (2011) 3054–3059. https://doi.org/10.1016/j.solmat.2011.06.032.

[9] A. Abiola-Ogedengbe, H. Hangan, K. Siddiqui, Experimental investigation of wind effects on a standalone photovoltaic (P.V.) module, Renew. Energy. 78 (2015) 657–665. https://doi.org/10.1016/j.renene.2015.01.037.

[10] C. Honsberg, S. Bowden, PVEducation, (n.d.). https://www.pveducation.org/pvcdrom/solar-cell-operation/solar-cell-structure (accessed June 29, 2021).

[11] C. Borri, M. Gagliardi, M. Paggi, Fatigue crack growth in silicon solar cells and hysteretic behaviour of busbars, Sol. Energy Mater. Sol. Cells. 181 (2018) 21–29. https://doi.org/10.1016/j.solmat.2018.02.016.

[12] A. Omazic, G. Oreski, M. Halwachs, G.C. Eder, C. Hirschl, L. Neumaier, G. Pinter, M. Erceg, Relation between degradation of polymeric components in crystalline silicon P.V. module and climatic conditions: A literature review, Sol. Energy Mater. Sol. Cells. 192 (2019) 123–133. https://doi.org/10.1016/j.solmat.2018.12.027.

[13] T.H. Kim, N.C. Park, D.H. Kim, The effect of moisture on the degradation mechanism of multi-crystalline silicon photovoltaic module, Microelectron. Reliab. 53 (2013) 1823–1827. https://doi.org/10.1016/j.microrel.2013.07.047.

[14] S. Kumar, R. Meena, R. Gupta, Imaging and micro-structural characterization of moisture induced degradation in crystalline silicon photovoltaic modules, Sol. Energy. 194 (2019) 903–912. https://doi.org/10.1016/j.solener.2019.11.037.

[15] H. Xiong, C. Gan, X. Yang, Z. Hu, H. Niu, J. Li, J. Si, P. Xing, X. Luo, Corrosion behavior of crystalline silicon solar cells, Microelectron. Reliab. 70 (2017) 49–58. https://doi.org/10.1016/j.microrel.2017.01.006.

[16] N.C. Park, W.W. Oh, D.H. Kim, Effect of temperature and humidity on the degradation rate of multicrystalline silicon photovoltaic module, Int. J. Photoenergy. 2013, pp. 1–10. https://doi.org/10.1155/2013/925280.

[17] M. Kumar, A. Kumar, Power estimation of photovoltaic system using 4 and 5-parameter solar cell models under real outdoor conditions, 2018 IEEE 7th World Conf. Photovolt. Energy Conversion, WCPEC 2018 - A Jt. Conf. 45th IEEE PVSC, 28th PVSEC 34th EU PVSEC. (2018) 721–726. https://doi.org/10.1109/PVSC.2018.8547765.

[18] M. Kumar, A. Kumar, Experimental characterization of the performance of different photovoltaic technologies on water bodies, Prog. Photovoltaics Res. Appl. 28 (2020) 25–48. https://doi.org/10.1002/pip.3204.

[19] M. Kumar, A. Kumar, R. Gupta, Comparative degradation analysis of different photovoltaic technologies on experimentally simulated water bodies and estimation of evaporation loss reduction, Prog. Photovoltaics Res. Appl. 29 (2021) 357–378. https://doi.org/10.1002/pip.3370.
[20] K. Tagawa, A. Kutani, P. Qinglin, Effect of sand erosion of glass surface on performances of photovoltaic module, Mech. Eng. Conf. Sustain. Res. Innov. (2012) 75–77.
[21] N.S. Beattie, R.S. Moir, C. Chacko, G. Buffoni, S.H. Roberts, N.M. Pearsall, Understanding the effects of sand and dust accumulation on photovoltaic modules, Renew. Energy. 48 (2012) 448–452. https://doi.org/10.1016/j.renene.2012.06.007.
[22] C. Buerhop, C.J. Brabec, S. Wirsching, S. Gehre, T. Pickel, T. Winkler, A. Bemm, J. Mergheim, C. Camus, J. Hauch, Lifetime and degradation of pre-damaged PV-modules – Field study and lab testing, PVSC (2017) 3500–3505. https://doi.org/10.1109/pvsc.2017.8366096.
[23] A.W. Czanderna, F.J. Pern, Encapsulation of P.V. modules using ethylene vinyl acetate copolymer as a pottant: A critical review, Sol. Energy Mater. Sol. Cells. 43 (1996) 101–181. https://doi.org/10.1016/0927-0248(95)00150-6.
[24] M. Kempe, Encapsulant Materials for P.V. Modules, Hoboken, NJ, USA: John Wiley & Sons, Ltd., 2016. https://doi.org/https://doi.org/10.1002/9781118927496.ch43.
[25] R. Meena, S. Kumar, R. Gupta, Comparative investigation and analysis of delaminated and discolored encapsulant degradation in crystalline silicon photovoltaic modules, Sol. Energy. 203 (2020) 114–122. https://doi.org/10.1016/j.solener.2020.04.041.
[26] A. Kraft, L. Labusch, T. Ensslen, I. Durr, J. Bartsch, M. Glatthaar, S. Glunz, H. Reinecke, Investigation of acetic acid corrosion impact on printed solar cell contacts, IEEE J. Photovoltaics. 5 (2015) 736–743. https://doi.org/10.1109/JPHOTOV.2015.2395146.
[27] M.D. Kempe, G.J. Jorgensen, K.M. Terwilliger, T.J. McMahon, C.E. Kennedy, T.T. Borek, Acetic acid production and glass transition concerns with ethylene-vinyl acetate used in photovoltaic devices, Sol. Energy Mater. Sol. Cells. 91 (2007) 315–329. https://doi.org/10.1016/j.solmat.2006.10.009.
[28] J.M. Kuitche, R. Pan, G. Tamizhmani, Investigation of dominant failure mode(s) for field-aged crystalline silicon P.V. modules under desert climatic conditions, IEEE J. Photovoltaics. 4 (2014) 814–826. https://doi.org/10.1109/JPHOTOV.2014.2308720.

[29] M.C.C. de Oliveira, A.S.A. Diniz Cardoso, M.M. Viana, V. de F.C. Lins, The causes and effects of degradation of encapsulant ethylene vinyl acetate copolymer (E.V.A.) in crystalline silicon photovoltaic modules: A review, Renew. Sustain. Energy Rev. 81 (2018) 2299–2317. https://doi.org/10.1016/j.rser.2017.06.039.

[30] J.H. Wohlgemuth, P. Hacke, N. Bosco, D.C. Miller, M.D. Kempe, S.R. Kurtz, Assessing the causes of encapsulant delamination in P.V. modules, 2017 IEEE 44th Photovolt. Spec. Conf. PVSC 2017, IEEE (2017) 1–6. https://doi.org/10.1109/PVSC.2017.8366601.

[31] C. Cuddihy, E., Coulbert, Electricity from Photovoltaic Solar Cells, Flat-Plate Solar Array Project, Jet Propuls. Lab. Publ. VII: Modul (n.d.). https://authors.library.caltech.edu/15043/1/JPLPub86-31volVII.pdf.

[32] E.E. Van Dyk, J.B. Chamel, A.R. Gxasheka, Investigation of delamination in an edge-defined film-fed growth photovoltaic module, Sol. Energy Mater. Sol. Cells. 88 (2005) 403–411. https://doi.org/10.1016/j.solmat.2004.12.004.

[33] C. Peike, S. Hoffmann, P. Hülsmann, B. Thaidigsmann, K.A. Weiß, M. Koehl, P. Bentz, Origin of damp-heat induced cell degradation, Sol. Energy Mater. Sol. Cells. 116 (2013) 49–54. https://doi.org/10.1016/j.solmat.2013.03.022.

[34] R. Meena, S. Kumar, R. Gupta, Investigation and analysis of chemical degradation in metallization and interconnects using electroluminescence imaging in crystalline silicon photovoltaic modules, Conf. Rec. IEEE Photovolt. Spec. Conf. 2020-June (2020) 2596–2599. https://doi.org/10.1109/PVSC45281.2020.9300539.

[35] S. Chattopadhyay, R. Dubey, V. Kuthanazhi, J.J. John, C.S. Solanki, A. Kottantharayil, B.M. Arora, K.L. Narasimhan, V. Kuber, J. Vasi, A. Kumar, O.S. Sastry, Visual degradation in field-aged crystalline silicon P.V. modules in India and correlation with electrical degradation, IEEE J. Photovoltaics. 4 (2014) 1470–1476. https://doi.org/10.1109/JPHOTOV.2014.2356717.

[36] X. Yao, L. Herrera, S. Ji, K. Zou, J. Wang, Characteristic study and time-domain discrete-wavelet-transform based hybrid detection of series D.C. arc faults, IEEE Trans. Power Electron. 29 (2014) 3103–3115. https://doi.org/10.1109/TPEL.2013.2273292.

[37] P. Rajput, G.N. Tiwari, O.S. Sastry, B. Bora, V. Sharma, Degradation of mono-crystalline photovoltaic modules after 22 years of outdoor exposure in the composite climate of India, Sol. Energy. 135 (2016) 786–795. https://doi.org/10.1016/j.solener.2016.06.047.

[38] I. Duerr, J. Bierbaum, J. Metzger, J. Richter, D. Philipp, Silver grid finger corrosion on snail track affected P.V. modules - Investigation on degradation products and mechanisms, Energy Procedia. 98 (2016) 74–85. https://doi.org/10.1016/j.egypro.2016.10.083.

[39] O. Breitenstein, J.P. Rakotoniaina, M.H. Al Rifai, M. Werner, Shunt types in crystalline silicon solar cells, Prog. Photovoltaics Res. Appl. 12 (2004) 529–538. https://doi.org/10.1002/pip.544.

[40] P. Somasundaran, R. Gupta, Influence of local shunting on the electrical performance in industrial silicon solar cells, Sol. Energy. 139 (2016) 581–590. https://doi.org/10.1016/j.solener.2016.10.020.

[41] M. Barbato, M. Meneghini, V. Giliberto, D. Giaffreda, P. Magnone, R. De Rose, C. Fiegna, G. Meneghesso, Effect of shunt resistance on the performance of mc-Silicon solar cells: A combined electro-optical and thermal investigation, Conf. Rec. IEEE Photovolt. Spec. Conf. (2011) 1241–1245. https://doi.org/10.1109/PVSC.2012.6317827.

[42] C. Buerhop, S. Wirsching, A. Bemm, T. Pickel, P. Hohmann, M. Nieß, C. Vodermayer, A. Huber, B. Glück, J. Mergheim, C. Camus, J. Hauch, C.J. Brabec, Evolution of cell cracks in PV-modules under field and laboratory conditions, Prog. Photovoltaics Res. Appl. 26 (2018) 261–272. https://doi.org/10.1002/pip.2975.

[43] M. Kntges, I. Kunze, S. Kajari-Schrder, X. Breitenmoser, B. Bjørneklett, The risk of power loss in crystalline silicon based photovoltaic modules due to micro-cracks, Sol. Energy Mater. Sol. Cells. 95 (2011) 1131–1137. https://doi.org/10.1016/j.solmat.2010.10.034.

[44] T. Cheng, M. Al-Soeidat, D.D.-C. Lu, V.G. Agelidis, Experimental study of P.V. strings affected by cracks, J. Eng. 2019 (2019) 5124–5128. https://doi.org/10.1049/joe.2018.9320.

[45] D. Lausch, V. Naumann, O. Breitenstein, J. Bauer, A. Graff, J. Bagdahn, C. Hagendorf, Potential-induced degradation (P.I.D.): Introduction of a novel test approach and explanation of increased depletion region recombination, IEEE J. Photovoltaics. 4 (2014) 834–840. https://doi.org/10.1109/JPHOTOV.2014.2300238.

[46] J. Li, Y.C. Shen, P. Hacke, M. Kempe, Electrochemical mechanisms of leakage-current-enhanced delamination and corrosion in Si photovoltaic modules, Sol. Energy Mater. Sol. Cells. 188 (2018) 273–279. https://doi.org/10.1016/j.solmat.2018.09.010.

[47] W. Luo, Y.S. Khoo, P. Hacke, V. Naumann, D. Lausch, S.P. Harvey, J.P. Singh, J. Chai, Y. Wang, A.G. Aberle, S. Ramakrishna, Potential-induced degradation in photovoltaic modules: A critical review,

Energy Environ. Sci. 10 (2017) 43–68. https://doi.org/10.1039/c6ee02271e.

[48] N. Kim, H. Kang, K.J. Hwang, C. Han, W.S. Hong, D. Kim, E. Lyu, H. Kim, Study on the degradation of different types of backsheets used in P.V. module under accelerated conditions, Sol. Energy Mater. Sol. Cells. 120 (2014) 543–548. https://doi.org/10.1016/j.solmat.2013.09.036.

[49] W. Gambogi, Y. Heta, K. Hashimoto, J. Kopchick, T. Felder, S. MacMaster, A. Bradley, B. Hamzavytehrany, L. Garreau-Iles, T. Aoki, K. Stika, T.J. Trout, T. Sample, A comparison of key P.V. backsheet and module performance from fielded module exposures and accelerated tests, IEEE J. Photovoltaics. 4 (2014) 935–941. https://doi.org/10.1109/JPHOTOV.2014.2305472.

[50] J. Tracy, W. Gambogi, T. Felder, L. Garreau-Iles, H. Hu, T.J. Trout, R. Khatri, X. Ji, Y. Heta, K.R. Choudhury, Survey of material degradation in globally fielded P.V. modules, Conf. Rec. IEEE Photovolt. Spec. Conf. (2019) 874–879. https://doi.org/10.1109/PVSC40753.2019.8981140.

[51] S.S. Chandel, M. Nagaraju Naik, V. Sharma, R. Chandel, Degradation analysis of 28 year field exposed mono-c-Si photovoltaic modules of a direct coupled solar water pumping system in western Himalayan region of India, Renew. Energy. 78 (2015) 193–202. https://doi.org/10.1016/j.renene.2015.01.015.

Chapter 9

Synthesis and Characterization of Botanical Dye-Sensitized Solar Cell (DSSC) Based on TiO_2 Using Capsicum Annuum and Coriandrum Sativum Extracts

Amit Shrivastava[1], Rupali Shrivastava[2], and Manoj Gupta[3]

[1]Poornima University, Jaipur, India
[2]Vivekananda Global University, Jaipur, India
[3]JECRC University, India
Corresponding authors: apasjpr@gmail.com, rupali.rec@gmail.com

Abstract

In this research work, biological dyes containing beta-carotene have been utilized for developing the botanical dye-sensitized solar cells (DSSC). The DSSC setup consisted of TiO_2 nanoparticles coated with dye molecules as a working electrode for experimental work. Synthesis of TiO_2 nanoparticles was carried out using the sol-gel method. Extracts of Capsicum Annuum and Coriandrum Sativum rich in beta carotenes act as natural sensitizers, and when used with TiO_2, they act as light photon catchers. A mixture of KI, acetonitrile, and iodine served the solid-state thin-film electrolyte function. UV-VIS spectrophotometer and X-ray diffractometer were used to learn the semiconductor thin films' optical and structural characteristics. *I–V* characteristics of the fabricated TiO_2 thin film were also studied. The efficiency of DSSC, fabricated with two different biological dyes, was found to be 1.6% for Capsicum Annuum and 1.03% for Coriandrum Sativum.

9.1 Introduction

Demand for electrical energy is multiplying quickly due to increased urbanization in the last decade. With constantly increasing pollution, the need for

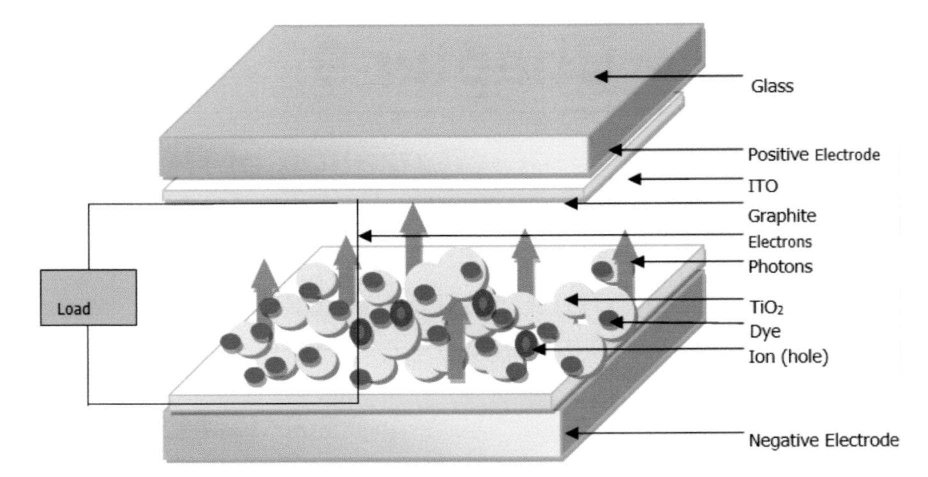

Figure 9.1 Section views of dye-sensitized solar cell (DSSC).

green and sustainable renewable energy sources is rising. Solar energy is one such renewable energy source that is eco-friendly, constant, and cost-effective; extensive research is being done on the development of solar cells [1–10, 28–30]. Dye-sensitized solar cells (DSSC) efficiently convert the sun's solar energy into electrical energy. Extensive research has been carried out on the advancement of the solar cell since its first development in 1991 by Grätzel *et al.* [11]. Since 1991, dye-sensitized solar cells (DSSC) have been utilized extensively as an alternative source of electrical energy due to associated advantages; being cost-effective, large area, and simple fabrication [12–16]. DSSC is a coupled redox system with electrolytes sandwiched between conductive glass-plates-based photo anode and cathode, as shown in Figure 9.1. Lost electrons from oxidized dye are regenerated from electrolyte redox couple. Photo anode comprises a plate of glass containing a coating of conductive oxide and a semiconductor material [17–19]. This research uses conductive oxide (indium doped tin oxide (ITO)) to coat a glass plate. The semiconductor material used is TiO_2 nanoparticles, which is a good platform for dye sensitizers to stand on. A counter cathode is also made of ITO covered glass plate with a thin film coating of carbon which acts as a catalyst for electrolyte reduction–oxidation reaction.

Such a DSSC's efficiency largely depends on the photon absorption capacity of the sensitized working electrode, photo anode in particular, which contributes to charge generation and its movement. Porous nature of TiO_2 nanoparticles with extensive bandgap and high binding energy results in superior adsorption of dye molecules on the surface of the photo anode. A larger number of dye molecules result in a greater amount of solar energy

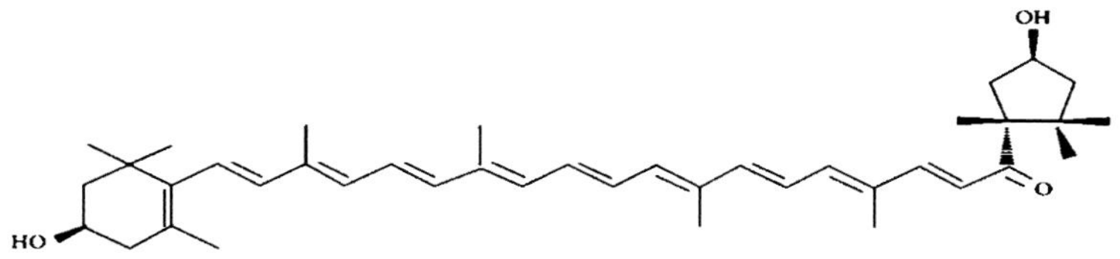

Figure 9.2 2D structure of Capsanthin.

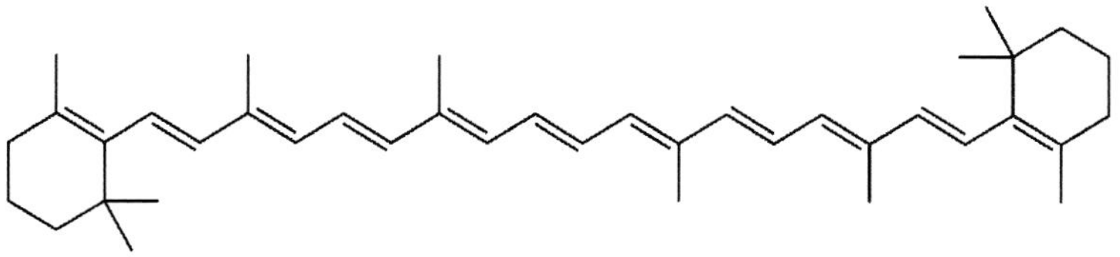

Figure 9.3 2D structure of β-carotenoid.

harvesting, which increases the efficiency of DSSC. Dyes get adsorbed on TiO_2 using active functional groups like carboxylic, azo, hydroxy, and carbonyl. A strongly bonded dye molecule can easily inject electrons into the conduction band of TiO_2. The dye-sensitized solar cell has good photon efficiency as per the literature review; thus, we selected such a system for our research work [1–7, 9–27].

Biological natural dyes extracted from diverse parts of plants are being extensively utilized in DSSC to reduce the cost involved and increase efficiency [8–12]. Besides chlorophyll, carotene is extensively used in DSSC and is a very good photocatalyst. Carotene is bright orange in color, with an absorption spectrum peak lying between 400 and 500 nm, which is the green/blue part of the spectrum [2–7, 5–20, 22–27]. In this research paper, extract of Capsicum Annuum and Coriandrum Sativum has been used as a source of carotene.

We have considered Capsicum Annuum for experimental study to have its bright red coloring property due to Capsanthin, C40H56O3, as a major carotenoid compound. As shown in the structure of Capsanthin given in Figure 9.2, it consists of a cyclopentane ring, a conjugated keto group, and 11 conjugated double bonds.

Coriandrum Sativum contains beta–carotene as a major constituent [16, 18], which is orange in color, but the orange color is masked due to the presence of green-colored chlorophyll. During the drying of coriander leaves, chlorophyll gets broken down, and the yellow/orange color of beta-carotenoid is visible. Beta-carotenoid $C_{40}H_{56}$ consists of eight isoprene units with cyclization at the ends of the long-conjugated chains, as shown in Figure 9.3.

9.2 Synthesis of TiO_2 Nanoparticles using Sol-Gel Technique

Titanium oxide nanoparticles were synthesized by a sol-gel technique using titanium tetra isopropoxide, double distilled water, and isopropanol as the raw material to initiate the synthesis. 100 ml of isopropanol was mixed with 15 ml of 1 M titanium tetra isopropoxide. The solution was stirred for 15 minutes using a magnetic stirrer. 10 ml of double-distilled water was mixed drop by drop in continents of the beaker. Then the mixture solution was shacked and mixed for 2 hours. A white precipitate was obtained, which was washed using distilled water after filtration. The obtained gel was kept for 24 hours and then dried with the help muffle furnace at 120°C. The dried TiO_2 is ground to a fine powder using a mortar and calcined at 350°C for 30 minutes in a muffle furnace. During the synthesis of TiO_2 nanoparticles, the following reactions occur according to the postulated mechanism:

$$Ti(OR)_3 + H_2O \rightarrow Ti(OH)_3 + R_2O \tag{9.1}$$

$$Ti(OH)_3 \rightarrow TiO_2 + H_2O \tag{9.2}$$

9.3 Preparation of Biological Dye Extract

For Capsicum Annuum and Coriandrum Sativum extract preparation, vegetables/leaves were crushed, and the obtained juice was filtered. To 60 ml of each extract, 30 ml of ethanol was added and allowed to boil for 30 minutes. After 30 minutes, the solid dye was obtained, which was allowed to cool and grind to a fine powder using mortar. 25 ml methanol, 5 ml acetic acids, and 15 ml of distilled water were added to the obtained fine powder. This mixture was then stirred at 450 rpm using a magnetic stirrer for one hour. Obtained solutions were filtered using Whatman filter paper 42 and stored separately in the freezer.

9.4 DSSC Fabrication

9.4.1 TiO_2 photo anode preparation

For the preparation of TiO_2-based photo anode, a layer of TiO_2 is deposited on indium-tin oxide (ITO), followed by adsorption of biological dye particles on the porous TiO_2 coating. The resistance of the ITO sheet is around 16 ohm/cm^2, but when the dye is applied, it results in increased sensitivity, and the photo anode, thus, becomes capable of absorbing even the lower energy photons [10, 17, 23, 27].

For application, TiO_2 paste was prepared by blending 4 g of synthesized TiO_2 nanoparticles, 3 ml of 0.1 M HNO_3, and 15 ml ethanol using a magnetic stirrer for 1 hour at 420 rpm. To control the thickness of TiO_2 film scotch tape is applied on the sides of ITO glass. Using dropper pipette and spin coating for 60 seconds at 1500 rpm, TiO_2 paste is spread on ITO. It is then sintered at 500°C for 1 hour and finally cooled to 50°C at a cooling rate 5°C/min. For the sensitizing process, absorption of dye into active areas of TiO_2 is needed; for this, dye solution is taken in a Petri dish, and TiO_2 coated ITO is dipped in the dye for around 1 hour. The plate is then cleaned with a mixture of distilled water and ethyl alcohol and dried in the open air.

9.4.2 Carbon coated photocathode preparation

The photocathode is prepared by coating the conducting side of the ITO plate with carbon using a graphite rod (soft graphite pencil). The coating is done using a soft graphite pencil by applying a light carbon film on the complete conductive side of the plate. Loose graphite particles are removed gently, and for long-lasting, the electrode was annealed at 500°C for 30 minutes and then finally washed with ethanol.

9.4.3 Electrolyte preparation

The electrolyte composition is KI 0.8 g and 0.13 g I_2 with 10 ml of acetonitrile. Mixing is done using a magnetic stirrer for an hour.

9.5 Results and Discussion

9.5.1 UV-Visible study of dye extract

To study the optical properties of the dye extracted and absorption behavior of TiO_2 nanoparticles, Perkin Elmer UV-Visible double beam spectrophotometer was used.

Absorption spectra of Capsicum Annuum were scanned in the range of 300–600 nm, as shown in Figure 9.4. The Capsicum Annuum extracts exhibit strong absorption broadband at 483 nm. This strong absorption band matches that of Capsanthin, which explains the photon capturing capacity of extract.

Absorption spectra of Coriandrum Sativum have also been scanned in the range of 300–600 nm, as shown in Figure 9.4. The absorption spectra of Coriandrum Sativum exhibit the highest absorption peak at 432 nm due to the presence of beta carotene in the extract. However, two less intense peaks

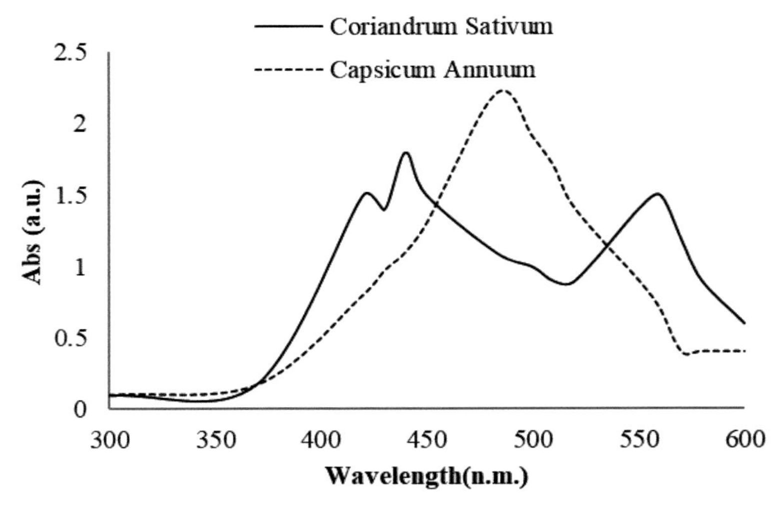

Figure 9.4 Adsorption curve of Capsicum Annuum and Coriandrum Sativum extract in 360–600 nm.

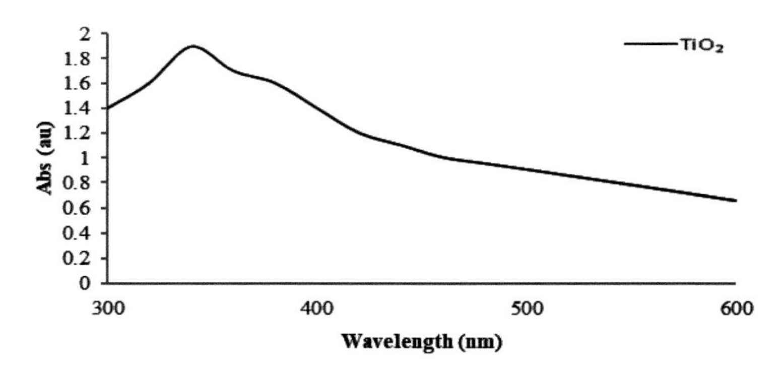

Figure 9.5 Adsorption curve of TiO_2 nanoparticles.

at 414 and 560 nm are also visible due to the presence of chlorophyll-a and b, respectively.

The study of the UV-visible behavior of TiO_2 nanoparticles is helpful in studying the bandgap and absorption behavior. On scanning in the range of 250–600 nm, the absorption peak maximum was obtained at 386 nm, as shown in Figure 9.5, which was further used to calculate the bandgap by the following equation:

$$E_g = hC/\lambda \tag{9.3}$$

where

E_g = Energy of bandgap
h = Planck's constant = 6.626×10^{-34} Joules/s
C = Speed of the light = 3×10^8 meter/s
λ = Sample absorbance

On calculating, the bandgap for TiO_2 nanoparticles is obtained as 3.4 eV.

9.6 Characterization of TiO_2 using XRD, FTIR, and SEM

The structural study of synthesized TiO_2 nanoparticles was done by X-ray diffraction (XRD) technique, and XRD spectra were recorded on PANalytical make X'Pert PRO MPD diffractometer (using CuK α radiation of $\lambda = 1.5418$ Å) within 2θ range of 10°–70° with a step of 0.1972°. Figure 9.6 shows the typical X-ray diffraction pattern of the synthesized TiO_2 nanoparticles. The XRD pattern exhibits the characteristics peaks at $2\theta°$ values of 25.2°, 25.8°, 37.2°, and 48°, confirming the anatase phase structure of TiO_2 nanoparticles and are in close agreement with previously reported XRD pattern of TiO_2 nanoparticles in literature. There is no additional diffraction peak corresponding to any other element that ratifies the synthesized sample's sanctity. The pointed and strong peaks suggest that the prepared nanoparticles are crystalline in nature.

The functional group of TiO_2 nanoparticles was identified by analyzing the FTIR. FTIR spectrum of TiO_2 is given in Figure 9.7. In this figure, the spectral peaks are at 3400 and 1631.78 cm^{-1} which are due to the bending and stretching of the –OH group. Peaks show the stretching and bending mode of

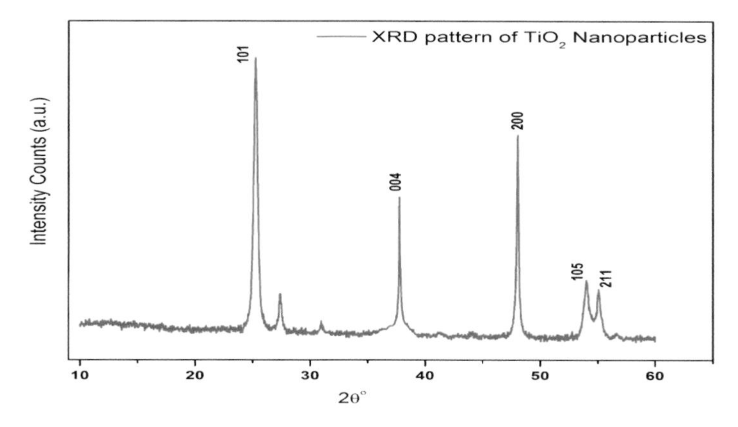

Figure 9.6 The X-ray diffraction patterns of synthesized TiO_2 nanoparticles.

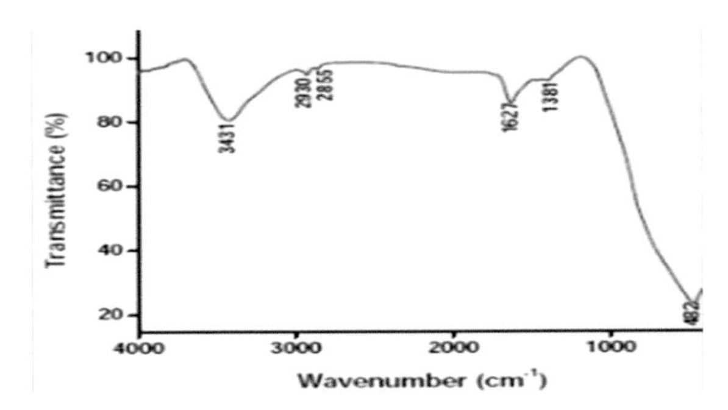

Figure 9.7 FTIR spectra of TiO_2 nanoparticles.

Figure 9.8 SEM images of synthesized TiO_2 nanoparticles.

Ti-O-Ti at 435.91 cm^{-1}, 466.77 cm^{-1} to 700 cm^{-1}. After carefully examining, no peak was found at 2900 cm^{-1}, which shows that after calcinations, all the natural compounds were removed from the samples.

To examine the shape and size of synthesized TiO_2 nanoparticles, categorization was done using scanning electron microscopy (SEM). SEM images at different magnifications are given in Figure 9.8. We initially crushed the synthesized TiO_2 powder in agate pestle mortar for 2 hours and ultra-sonicate for 30 minutes in 10-ml acetone solution. In the next step, we prepare a thin film of TiO_2 powder on a clean glass substrate using the spin coating technique. SEM images (a) and (b) clearly exhibit irregular distributed TiO_2 particles in an uneven cluster of TiO_2 particles. By further examination at higher magnification, it has been observed that most of the TiO_2 particles are spherical and possess a smooth surface. The average particle size of TiO_2 particles is observed to be ≈ 20–40 nm.

9.6.1 Efficiency calculation of DSSC

The efficiency (η) of a DSSC is given by the following relation [2]:

$$\text{Efficiency} = \text{Solar cell output power/Solar cell input power} \quad (9.4)$$

and per unit area power is calculated by the following relation [24]:

$$\text{Power per unit area} = (\text{Voltage at open circuit} \times \text{current at the short circuit})/\text{Area}. \quad (9.5)$$

Calculated data of efficiency of Capsicum Annuum and Coriandrum Sativum are summarized in Table 9.1.

9.6.1.1 Performance Analysis of DSSC

The performance of DSSC is measured by plotting the current–voltage curve. For measurement, current–voltage meter Keithley 2620A was used with sunlight as a light source. Measured current–voltage for Capsicum Annuum and Coriandrum Sativum are summarized in Table 9.1.

Using recorded DSSC, the time–voltage (Figure 9.9) and time–current curve (Figure 9.10) for both the dye extracts – Capsicum Annuum and Coriandrum Sativum – were plotted as shown in Figures. 9.9 and 9.10. The curve shows that voltage is more for Coriandrum Sativum dye than Capsicum Annuum dye. The curves also show that current flow was maximum when the intensity of sunlight is maximum at noon, and, after that, it starts decreasing.

Table 9.1 Efficiency of DSSC for Capsicum Annuum and Coriandrum Sativum.

Dyes	**Molecular structure**	V_{OC} **(mV)**	I_{SC} **(mA)**	A **(cm²)**	P_{IN} **(mW/cm²)**	η **(%)**
Capsicum Annuum	Capsanthin	0.38	5.15	2	61.25	1.6
Coriandrum Sativum	Beta Carotne	0.52	2.42	2	61.25	1.03

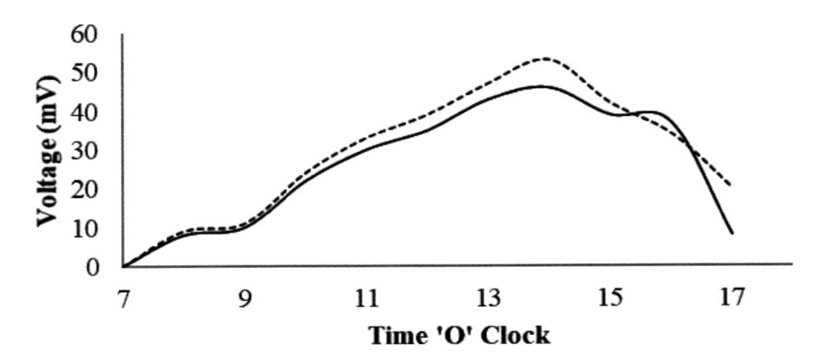

Figure 9.9 Time–voltage curve of DSSC.

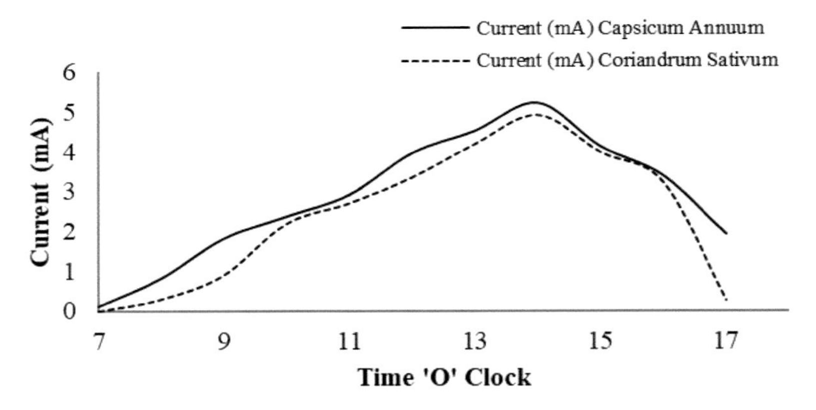

Figure 9.10 Time–current curve of DSSC.

9.7 Conclusion

Synthesis of TiO_2 nanoparticles was successfully achieved, as proved from XRD, SEM, and FTIR studies. Photosensitization activity and its variation with time for two natural dye extracts were compared. On analyzing the data of the time–current curve, more voltage was obtained in the case of Coriandrum Sativum dye compared to Capsicum Annuum dye. The curves also show the direct relationship between current flow and the intensity of sunlight. Thus, it can be concluded that Capsicum Annuum DSSC is more efficient than Coriandrum Sativum dye. Results obtained report Capsicum Annuum dye as a new sensitizer used in solar cells.

References

[1] International Telecommunication Union (ITU), Report on Climate Change, Oct. 2008.

[2] G. Koutitas and P. Demestichas, 'A review of energy efficiency in telecommunication networks', Proceedings of Telecommunications Forum (TELFOR), Serbia, Nov. 2009, pp. 1-4.

[3] Gartner Report, Financial Times, 2007.

[4] I. Cerutti, L. Valcarenghi, and P. Castoldi, 'Designing power-efficient WDM ring networks,' ICST International Conference on Networks for Grid Applications, Athens, 2009, pp. 101–108.

[5] W. Vereecken, *et al.*, 'Energy efficiency in thin client solutions,' ICST International Conference on Networks for Grid Applications, Athens, 2009, pp. 109–116.

[6] J. Haas, T. Pierce, and E. Schutter, 'Datacenter design guide,' Whitepaper, The Greengrid, 2009.

[7] Intel, 'Turning challenges into opportunities in the data center,' White Paper.

[8] Z. N. Kayani, Maria, S. Riaz, and S. Naseem, 'Magnetic and antibacterial studies of sol-gel dip coated Ce doped TiO2 thin films: Influence of Ce contents,' Ceram. Int., vol. 46, no. 1, pp. 381–390, 2020.

[9] S. K. Das, S. Ganguli, H. Kabir, J. I. Khandaker, and F. Ahmed, 'Performance of natural dyes in dye-sensitized solar cell as photosensitizer,' Trans. Electr. Electron. Mater., vol. 21, no. 1, pp. 105–116, 2020.

[10] K. Mensah-Darkwa, F. O. Agyemang, D. Yeboah, and S. Akromah, 'Dye-sensitized solar cells based on graphene oxide and natural plant dye extract,' Mater. Today Proc., 2020, Pages 107–161.

[11] G. Calogero and A. Bartolotta, 'Chapter four - dye sensitized solar cells: from synthetic dyes to natural pigments,' in *Solar Cells and Light Management: Materials, Strategies and Sustainability*. Amsterdam, Netherlands: Elsevier, 107-161, 2020.

[12] A. M. Ammar, H. S. H. Mohamed, M. M. K. Yousef, G. M. Abdel-Hafez, A. S. Hassanien, and A. S. G. Khalil, 'Dye-sensitized solar cells (DSSCs) based on extracted natural dyes,' J. Nanomater., vol. 2019, pp. 1–11.

[13] P. Trihutomo, S. Soeparman, D. Widhiyanuriyawan, and L. Yuliati, 'Performance improvement of dye-sensitized solar cell- (DSSC-) based natural dyes by Clathrin protein,' Int. J. Photoenergy, vol. 2019, pp. 1–10.

[14] A. Hayat, A. E. E. Putra, N. Amaliyah, and S. S. Pandey, 'Clitoria ternatea flower as natural dyes for dye-sensitized solar cells,' IOP Conf. Ser. Mater. Sci. Eng., vol. 619, no. 1, 2019, pp. 1–6.
[15] D. K. Doda, A. Shrivastava, and M. Bundele, 'Design & optimization of a power generation system based on renewable energy technologies,' Int. J. Innov. Technol. Explor. Eng., vol. 8, no. 11, pp. 1134–1138, 2019.
[16] M. J. García-Salinas and M. J. Ariza, 'Optimizing a simple natural dye production method for dye-sensitized solar cells: Examples for betalain (bougainvillea and beetroot extracts) and anthocyanin dyes,' Appl. Sci., vol. 9, no. 12, 2019, pp. 1–20.
[17] K. S. Pawar, P. K. Baviskar, Inamuddin, A. B. Nadaf, S. Salunke-Gawali, and H. M. Pathan, 'Layer-by-layer deposition of TiO2–ZrO2 electrode sensitized with Pandan leaves: Natural dye-sensitized solar cell,' Mater. Renew. Sustain. Energy, vol. 8, no. 2, pp. 1–9, 2019.
[18] B. O'Regan and M. Grätzel, 'A low-cost, high-efficiency solar cell based on dye-sensitized colloidal TiO2 films,' Nature, vol. 353, pp. 737–740, 1991.
[19] N. N. Rosli, M. A. Ibrahim, N. Ahmad Ludin, M. A. Mat Teridi, and K. Sopian, 'A review of graphene based transparent conducting films for use in solar photovoltaic applications,' Renew. Sustain. Energy Rev., vol. 99, no. May 2017, pp. 83–99, 2019.
[20] M. J. Yun, Y. H. Sim, S. I. Cha, and D. Y. Lee, 'Leaf anatomy and 3-D structure mimic to solar cells with light trapping and 3-D arrayed submodule for enhanced electricity production,' Sci. Rep., vol. 9, no. 1, pp. 1–9, 2019.
[21] M. B. Shitta, E. O. B. Ogedengbe, O. B. Familoni, and O. T. Ogundipe, 'Diffusion characterization and modelling of Mimosa pudica extract towards the production of organic solar module,' AIAA Propulsion and Energy Forum Exposition, 2019, pp. 1–14.
[22] F. Nurosyid, D. D. Pratiwi, and K. Kusumandari, 'The effect of immersion temperature using chlorophyll sensitizer (Amaranthus hybridus L.) on the performance of dye-sensitized solar cells,' J. Phys. Conf. Ser., vol. 1397, no. 1, 2019, pp. 1–6.
[23] C. Cari, K. Khairuddin, T. Y. Septiawan, P. M. Suciatmoko, D. Kurniawan, and A. Supriyanto, 'The preparation of natural dye for dye-sensitized solar cell (DSSC),' AIP Conference Proceedings, vol. 2014, September 2018, pp. 1–5.
[24] J. Liu, Y. Li, S. Arumugam, J. Tudor, and S. Beeby, 'Investigation of low temperature processed titanium dioxide (TiO2) films for printed

dye sensitized solar cells (DSSCs) for large area flexible applications,' Mater. Today Proc., vol. 5, no. 5, pp. 13846–13854, 2018.

[25] K. Sharma, V. Sharma, and S. S. Sharma, 'Dye-sensitized solar cells: Fundamentals and current status,' Nanoscale Res. Lett., vol. 13, 2018, pp. 1–46.

[26] R. Shrivastava and M. Meena, 'Spectrophotometric determination of stability constant of naphthol green-B dye using Ni(II) as an analytical reagent,' Adv. Sci. Engg. Medicine, vol. 11, no. 1, pp. 69-73, 2019.

[27] D. Tahir, W. Satriani, P. L. Gareso, and B. Abdullah, 'Dye sensitized solar cell (DSSC) with natural dyes extracted from Jatropha leaves and purple Chrysanthemum flowers as sensitizer,' J. Phys. Conf. Ser., vol. 979, The 2nd International Conference on Science (ICOS), Makassar, Indonesia, 2017, pp. 2–3.

[28] W. Ghann, H. Kang, T. Sheikh, et al., Fabrication, Optimization and Characterization of Natural Dye Sensitized Solar Cell. Sci Rep 7, 41470 (2017). DOI: 10.1038/srep41470.

[29] C. P. Lee, C. T. Li, and K. C. Ho, 'Use of organic materials in dye-sensitized solar cells,' Mater. Today, vol. 20, no. 5, pp. 267–283, 2017.

[30] N. Jamalullail, I. S. Mohamad, M. N. Norizan, N. A. Baharum, and N. Mahmed, 'Short review: Natural pigments photosensitizer for dye-sensitized solar cell (DSSC),' 2017 IEEE 15th Student Conference on Research and Development (SCOReD), Putrajaya, 2017, pp. 344-349.

[31] T. V. S. S. P. Sashank, B. Manikanta, and A. Pasula, 'Fabrication and experimental investigation on dye sensitized solar cells using titanium dioxide nano particles,' Mater. Today Proc., vol. 4, no. 2, pp. 3918–3925, 2017.

[32] O. Adedokun, K. Titilope, and A. O. Awodugba, 'Review on natural dye-sensitized solar cells (DSSCs),' Int. J. Eng. Technol. IJET, vol. 2, no. 2, p. 34, 2016, pp. 34–41.

[33] M. Bernardi, M. Palummo, C. Grossman, and R. Scienti, 'Extraordinary sunlight absorption and one nanometer thick photovoltaics using two-dimensional monolayer materials,' Nano Lett., vol. 13, no. 8, pp. 3664–3670, 2013.

[34] H. Alarcón, G. Boschloo, P. Mendoza, J. L. Solis, and A. Hagfeldt, 'Dye-sensitized solar cells based on nanocrystalline TiO2 films surface treated with Al3+ ions: Photovoltage and electron transport studies,' J. Phys. Chem. B, vol. 109, no. 39, pp. 18483–18490, 2005.

[35] A. Shrivastava, D. K. Doda, and M. Bundele, 'Economic and environmental impact analysis of hybrid generation system in respect to Rajasthan', Environ. Sci. Pollution Res., vol. 28, pp. 3906–3912, July 2020.

[36] D. K. Doda, A. Shrivastava, and M. Bundele, 'Design and optimization of a power generation system based on renewable energy technologies', Int. J. Innovative Technol. Exploring Engg., vol. 8, no. 11, pp. 1134-1138, September 2019.

[37] D. K. Doda, M. Bundele, and A. Shrivastava, 'Financial feasibility of a remote village cluster without access to electricity' Int. J. Environ. Sustainable Develop., vol. 20, no. ¾, 2021 pp. 354–365.

Index

R

S

X

About the Editors

Gaurav Saini, is presently working as the Professor (Assistant) in the Department of Mechanical Engineering, Harcourt Butler Technical University Kanpur, Uttar Pradesh. Dr. Saini has Post-Doctoral Fellow experience with the Department of Sustainable Energy Engineering, Indian Institute of Technology Kanpur. Prior to joining IIT Kanpur, Dr. Saini was serving as the Professor (Assistant) in the School of Advanced Materials, Green Energy and Sensor Systems, Indian Institute of Engineering Science and Technology (IIEST) Shibpur India. He received his Ph.D. in Turbomachines (Hydrokinetic Turbines) in the year 2020 and M. Tech, (Fluid Machinery and Energy Systems) in the year 2014 from Indian Institute of Technology Roorkee, Uttarakhand India.

After his Ph.D. from IIT Roorkee, he was working as Project Fellow in the Department of Hydro and Renewable Energy, Indian Institute of Technology Roorkee. His research areas include Renewable Energy (Hydrokinetic Energy, Wind Power, and Biomass), Computational Fluid Dynamics (CFD) and Fluid Mechanics and Turbomachines; Fluid Power. He has published several research publications on renewable energy technologies in different international journals of repute. He has also presented his research at different international and national platforms and he received accolades from various peers working in the same area across the globe. Dr. Gaurav is skilled in Computational Fluid Dynamics (CFD) - numerical modelling and rotodynamics analysis, Multiphase flow analysis, Modeling of various renewable energy resources viz. wind, marine, solar and hydrokinetic energy for rural applications, wind and hydrokinetic- Technology selection and design, Installation strategies, Performance evaluation and O&M issues.

Korhan Cengiz, received the Ph.D. degree in electronics engineering from Kadir Has University, Turkey, in 2016. Since 2021, he has been the Department Vice Chair of the Electrical-Electronics Engineering, Trakya University, Turkey. He is the author of over 40 articles in publications including *IEEE Internet of Things Journal* and *IEEE Access*, four international patents, and four book chapters. His research interests include WSNs, indoor positioning systems, IoT, and 5G. He is a Handling Editor for *Microprocessors and Microsystems*, Elsevier, and an Associate Editor for *IET Electronics Letters*. He has served as a Guest Editor for *IEEE Internet of Things Magazine*.

Sesha Srinivasan, is presently working as a Professor with the Florida Polytechnique University. Prior to that, he spent five years as a tenure-track Assistant Professor of physics with Tuskegee University. Srinivasan has more than a decade of research experience in the interdisciplinary areas of solid state and condensed matter physics, inorganic chemistry, chemical, and materials science engineering. His doctorate focused on the development of various rare-earth, transition metals and intermetallic alloys, composites, nanoparticles, and complex hydrides for reversible hydrogen storage applications. He and his post-doctoral advisor have collaborated extensively with scientists around the world for the hydrogen storage on lightweight complex hydrides, which was funded by the U.S. Department of Energy (DOE) and the World Energy Network (WE-NET), Japan.

His career has included two years at the University of Hawaii followed by six years at the University of South Florida where he was a Research Scientist with the Clean Energy Research Center (CERC) under the leadership of Prof. Elias Stefanakos and Prof. Yogi Goswami. He has also served as an Associate Director of the Florida Energy Systems Consortium (FESC) where he coordinated a number of research projects on clean energy and the environment, which were funded by a $9 million grant from the Florida Office of Energy.

Srinivasan has been awarded several research grants worth more than $1 million from both federal (DOE, National Science Foundation, and Office of Naval Research) and private (BP-Oil Spill and QuantumSphere Inc.) funding sources. He was recently awarded two U.S. patents on hydrogen storage nanomaterials development and methodologies, and two U.S. patents are pending. He has published six book chapters and review articles, more than 100 journal publications, and many more peer-reviewed conference proceedings.

Sanjeevikumar Padmanaban, (Member'12–Senior Member'15, IEEE) received the Ph.D. degree in electrical engineering from the University of Bologna, Bologna, Italy, in 2012. He was an Associate Professor with VIT University from 2012 to 2013. In 2013, he joined the National Institute of Technology, India, as a Faculty Member. In 2014, he was invited as a Visiting Researcher at the Department of Electrical Engineering, Qatar University, Doha, Qatar, funded by the Qatar National Research Foundation

(Government of Qatar). He continued his research activities with the Dublin Institute of Technology, Dublin, Ireland, in 2014. Further, he served as an Associate Professor with the Department of Electrical and Electronics Engineering, University of Johannesburg, Johannesburg, South Africa, from 2016 to 2018. Since 2018, he has been a Faculty Member with the Department of Energy Technology, Aalborg University, Esbjerg, Denmark. He has authored over 300 scientific papers. S. Padmanaban was the recipient of the Best Paper cum Most Excellence Research Paper Award from IET-SEISCON'13, IET-CEAT'16, IEEE-EECSI'19, IEEE-CENCON'19, and five best paper awards from ETAEERE'16 sponsored Lecture Notes in Electrical Engineering, Springer book. He is a Fellow of the Institution of Engineers, India, the Institution of Electronics and Telecommunication Engineers, India, and the Institution of Engineering and Technology, U.K. He is an Editor/Associate Editor/Editorial Board for refereed journals, in particular the *IEEE Systems Journal*, IEEE Transactions on Industry Applications, *IEEE Access*, *IET Power Electronics*, *IET Electronics Letters*, and *Wiley-International Transactions on Electrical Energy Systems*, Subject Editorial Board Member—Energy Sources—*Energies* journal, MDPI, and the Subject Editor for the *IET Renewable Power Generation*, *IET Generation, Transmission and Distribution*, and *FACTS Journal* (Canada)

Krishna Kumar, is presently working as a Research and Development Engineer with UJVN Ltd. Before joining UJVNL. He worked as an Assistant Professor with BTKIT, Dwarahat. He received the B.E. degree in electronics and communication engineering from Govind Ballabh Pant Engineering College, Pauri Garhwal, and the M.Tech degree in digital systems from Motilal Nehru NIT Allahabad. He is currently working toward the Ph.D. degree from the Indian Institute of Technology, Roorkee, India. He has over 11 years of experience and has published numerous research papers in international journals. His research areas include renewable energy and artificial intelligence.